# RACE
# UNMASKED

—

# MICHAEL YUDELL

# RACE UNMASKED

———

Biology and Race

in the Twentieth Century

**FOREWORD BY J. CRAIG VENTER**

COLUMBIA UNIVERSITY PRESS
NEW YORK

Columbia University Press
*Publishers Since 1893*
New York   Chichester, West Sussex
cup.columbia.edu

Library of Congress Cataloging-in-Publication Data
Yudell, Michael.
Race unmasked : biology and race in the twentieth century / Michael Yudell.
pages   cm
Includes bibliographical references and index.
ISBN 978-0-231-16874-8 (cloth : alk. paper) — ISBN 978-0-231-53799-5 (electronic)
1. Eugenics—History—20th century. 2. Race—History—20th century. 3. Human biology—History—20th century. I. Title.

HQ751.Y83   2014
363.9'2009'04—dc23

2013043152

Columbia University Press books are printed on permanent and durable acid-free paper.
This book is printed on paper with recycled content.
Printed in the United States of America

c 10 9 8 7 6 5 4 3 2 1

Cover design: Mary Ann Smith

References to websites (URLs) were accurate at the time of writing. Neither the author nor Columbia University Press is responsible for URLs that may have expired or changed since the manuscript was prepared.

*To Gerald Gill,*
*for all that you gave*

# CONTENTS

# FOREWORD

## J. CRAIG VENTER, PH.D.

The concept of race is a deeply embedded historical challenge for human societies, one that Michael Yudell clearly illustrates in this excellent book. In my research on the human genome, we have also found race to be a social construct, not a scientific one. Despite the many claims otherwise, science and scientists *are not* infallible or unbiased when it comes to conceptualizing race. After all, as this book shows, some very notable scientists, even some from recent history, have espoused "scientific theories" to support their racial beliefs. Classification of species has been a part of science for centuries; thus scientists have struggled with these ideas for centuries.

Only recently, however, have we learned to measure the genetic code quantitatively and acknowledged that most of the previous classification was based primarily on visual differences. This type of behavior is clearly a human trait. We like people who look like us. We view the "same" as safe and reinforcing, and difference as foreign and potentially dangerous. One of the many ironies of this overly simplistic, crude classification system is that at some point in human history there may have been a selective advantage for being wary of the potentially disease-carrying stranger coming to your cave, village, or town. Perhaps that is one reason such narcissistic genetic traits are with us today. Yet despite these tendencies toward self-liking, our genomes show evidence of extensive interbreeding going back tens of thousands of years.

One of the reasons I moved into molecular biology and genomics is that it is a quantitative field. You either have the DNA sequence or you don't. It is either accurately measured or it is not. And the discovery that our genetic code carries within each of us our entire genetic history, as well as human evolutionary history, permits for the first time a quantitative basis for deciphering our history, our evolution, and our similarities with and differences from each other.

All modern humans originated in Africa, but some Africans migrated out 300,000 to 400,000 years ago and evolved into the Neanderthals. Not that long ago, it was a hotly debated question whether Neanderthals and modern humans interbred. Based on the incredible work of Svante Paabo and colleagues sequencing several Neanderthal genomes from bone fragments found in the Vindija Cave in Croatia, we now know that a group of modern humans that migrated out of Africa interbred with Neanderthals between 40,000 and 90,000 years ago. Paabo's findings show that 1 to 4 percent of the genomes from East Asians and Europeans are of Neanderthal origin. Neanderthal genes contributed changes in skin and hair that perhaps helped these populations adapt to colder climates.

The Neanderthal–modern human interactions are only a single example of how human populations have interacted and intermixed throughout history. Hellenthal and colleagues reported this year in the journal *Science* (343 [2014]: 747–751) that admixture has been "an almost universal force shaping human populations." Their work also highlights the impact of the Mongol empire, the Arab slave trade, and the Bantu expansion in influencing humanity's genetic code.

Our genomes have been mixed and remixed with every generation, so much so that the notion of any "pure" human population is absurd. In my talk at the White House in 2000 to announce the completion of the first draft of the human genome, I said that race has no basis in the genetic code. The results of genome sequencing over the last thirteen years only prove my point more clearly. It is a fact that there are greater genetic differences between individuals of the same "racial" group than between individuals of different groups. The problem we face with the emergence of so-called race-based medicine is the same problem as with applying average or "normal" clinical values to any individual. They just don't work well.

Genomics is about understanding that the uniqueness of each and every one of us cannot be determined by the broad general groups to which we appear to belong. Generalizations might work for clinical values and for populations, but not at the individual level. As human genome sequencing becomes a standard part of clinical practice over the next few decades, we will find that we all belong to a multitude of different human populations in terms of disease risk, drug responses, and differences in drug metabolism. Some of these differences might date back to the Neanderthal–modern human mating that increased the risk for Type 2 diabetes over that of people who came out of Africa much later. Or they could derive from ancestors from Bedouin tribes whose lifestyle selected for Type 2 diabetes as a survival advantage.

We know that with all the admixture that has occurred throughout human history, skin color will not predict what will be found in your genetic code. Race and race concepts will not stand the test of time. The socially driven attitudes and biases that have for too long found a home in science will slowly be replaced. Unfortunately, in the case of the scientific race concept this has indeed happened slowly.

# ACKNOWLEDGMENTS

It may be cliché, but it is nonetheless true that a book is much more than the singular effort of its author. In the case of this book, I can confirm this—I could not have completed this work without the nurturing, encouragement, patience, and assistance of many good friends, colleagues, and family members.

I feel like I began this project in a different lifetime. Indeed, it was a long time ago that the inspiration for this book began at the City University of New York Graduate Center in a class on the history of public health with David Rosner and Gerald Markowitz. David has played a unique role in my intellectual life. He guided me through too many years of graduate school, all the while remaining a patient and dedicated mentor who encouraged me every step of the way. From David I learned that not simply does history matter, but that matters of history can be the basis for living an intensely political, satisfying, and moral life.

One of David's most important contributions to my career was introducing me to Rob DeSalle at the American Museum of Natural History, and Rob and I have been close friends and collaborators ever since. Rob invited me to join his molecular laboratory as a student (in conjunction first with my studies at CUNY and later at Columbia), and we made a deal; I would bring to him history of science, medicine, and public health texts that we would read together and he would teach me genetics. From that beginning, Rob and I would go on to write two books together, with a third on the way, and my years in his lab helped me develop into the scholar I am today.

So much of the conceptual framework of this book was developed in conversation with friends both inside and outside academia. Kelvin Sealey not only read the entire text and offered his careful edits but he also provided both intellectual and emotional support during this long process. Kelvin and I started as graduate students together at the Graduate Center, CUNY, and together migrated to separate departments at Columbia. Over the course of that time we have supported one another in both friendship and intellectual pursuits.

James Colgrove, with whom I began my studies at Columbia and who is now a colleague and collaborator, read and discussed parts of the book and deserves special thanks. Joanna Radin, who is herself emerging as one of the most thoughtful historians of science, now at Yale, read the entire manuscript and helped me think through some of the challenges inherent to this topic. Her detailed comments made this a richer book.

Others, including Avi Patt, Bette Begleiter, Paul Messing, Bill Shein, Elizabeth Robilotti, Neil Schwartz, Cindy Lobel, Terrence Kissack, Tracy Morgan, Ariel De, Howard Rosenbaum, Michael Russello, Rick Baker, James Bonacum, Jorge Brito, Stuart Zicherman, Greg Moss, the late Myra Frazier, Jonathan Mannina, Sandy Kandel, and Seth Krevat, have, at various times, been forced to discuss or read the material herein and deserve my gratitude.

Many of the ideas and impulses in this book can be traced back to Colin Palmer, who as an early mentor pushed me to consider not just the idea of race in science but also the relationship of that idea to both the lives of African Americans and to African American history. While I am not satisfied that this project does enough on both counts, this is a better book because of him.

Others still were incredibly generous with their time and advice on the manuscript, including Keith Wailoo, Richard Lewontin, Susan Reverby, Ruth Schwartz Cowan, Merlin Chowkwanyun, Arthur Caplan, Janet Golden, Richard Sharp, and David Barnes. Additionally, Amy Fairchild, Gerald Markowitz and Elizabeth Blackmar read the entire manuscript and made extensive comments in its dissertation stage. Finally, very special thanks to Dr. J. Craig Venter for taking time to write the foreword to this book and for his work challenging scientists to reconsider their use of race as a variable in research.

To the many librarians and archivists who helped me along the way I cannot say enough thank-yous. Rob Cox, formerly the chief librarian at the American Philosophical Society and currently the head of the Special Collections and University Archives at the University of Massachusetts, Amherst, was my guide to the collections at both repositories. Rob's keen insight into the history of genetics and eugenics was invaluable. A Library Resident Research Fellowship from the American Philosophical Society during the summer of 2004 provided me with the resources to complete most of the primary source research for this project. At the society Valerie-Ann Lutz, Joseph-James Ahern, Roy Goodman, Charles Greifenstein, and the entire staff provided invaluable assistance to me and my work. The librarians at the American Museum of Natural History library and archives offered their careful assistance to this project. Finally, Leonard Bruno, science manuscript historian at the Library of Congress in Washington, D.C., helped me navigate the as yet unprocessed E. O. Wilson papers, and the staff at the Stanford University Archives helped me navigate the then unprocessed Stephen Jay Gould papers.

My colleagues at the Drexel University School of Public Health have been supportive as I have worked to complete this book. My former dean, Marla Gold, my former and current department chairs Lisa Ulmer and Ann Klassen helped me carve out time in a busy schedule of teaching and other research responsibilities to complete the manuscript. I also owe special thanks to the now nine years of students who have heard me lecture on this subject and whose thoughtful reactions to this material forced me to think more carefully about it. Aaron Pankiewicz, Geoffrey Vargish, Jamie Earnest, Nicole Gidaya, Deb Langer, John Donovan, and Lilliam Ambroggio were especially helpful. Extra special thanks to Phoebe Jones, whose editing skills and insight were invaluable as I completed the book. Finally, thanks to several current and former colleagues who read and talked about sections of the book: John Rich, John Rossi, Craig Newschaffer, Lisa Bowleg, Randall Sell, Augusta Villanueva, Marcus Kolb, and Hernando Perez.

I am grateful to have worked with Patrick Fitzgerald and his team at Columbia University Press, including Kathryn Schell, Bridget Flannery-McCoy, Leslie Kriesel, and Mike Ashby. Patrick was a wonderful editor, working closely with me every step of the way, and I am thankful for all that he did as he pushed me and guided me to make this a better book.

My mother and late father, Jane and Allen Yudell, instilled in me the progressive values that are at the core of my professional goals, and they deserve my deepest gratitude and love. My sister, Andrea Yudell-Nandi, has always been a loving friend in our journey through life. And my in-laws, Alan Rick and Debra Sacks, who came into my life in the middle of this long project, have offered only their deepest support.

My wife, Jacqueline Rick, whom I met on the downtown 1 train in New York City while we were both doctoral students at Columbia, is the center of my life, and this work was driven as much by her interest in my ideas as it was by her insistence that I finish the damn thing. Thank-you is not enough for her. Only my dedication to her as a husband and father of our daughters can begin to account for all that she has given me.

This book is dedicated to my mentor and friend, the late Gerald Gill. It was Gerald who inspired me (and several generations of undergraduates at Tufts University) to dive headfirst into the past in his seminars on the civil rights movement, the history of the American South, and African American history. Gerald was inspirational in the way he embraced the past. He did so with rigorous scholarly resolve and basic human decency with the hope of carving out a better future. This, along with his wry sense of humor and party trick–like encyclopedic recall, earned him the love and respect of his students, colleagues, and friends. For me, I saw the way Gerald lived his life as a scholar and teacher, and I wanted to be like him. I hope this book is another step in that direction.

# RACE
# UNMASKED

——

# INTRODUCTION

R ace, while drawn from the visual cues of human diversity, is an idea with a measurable past, identifiable present, and uncertain future. The concept of race has been at the center of both triumphs and tragedies in American history and has had an unmistakable impact on the human experience. It is a term used both casually and scientifically; a way people and groups choose to describe themselves and their ancestors; a way scientists and societies have chosen to describe and interpret the complexity that is human diversity and difference; and a way that doctors and public health officials make decisions about our health, both individually and collectively. It can be a source of pride, self-understanding, and resistance. Also of oppression and carnage. It is indeed an idea that has shaped the dreams and lives of generations.

This book tells the history of the formulation and preservation of the race concept and explores the role that science, particularly genetics and related biological disciplines, played in the making of America's racial calculus over the course of the twentieth century. In so doing, it shows where commonly held beliefs about the scientific nature of racial differences come from and examines the origins of the modern idea of race. The book also examines how ideas about race developed into a biological concept during the twentieth century, and how that concept has persisted in various incarnations as accepted scientific fact into the twenty-first. This is not, however, a story of the triumph of rational science over ignorance and racism. Instead, this book considers how this history shaped a contemporary paradox in thinking about the biological race concept; that is, that race can be understood to be both a critical methodological tool for biologists to make sense of human genetic diversity and, at the same time, widely believed *not* to be a particularly accurate marker for measuring that diversity.

The race concept in biology can be traced to eighteenth- and nineteenth-century debates about slavery, colonialism, and the nature of citizenship, which were driven by the sciences of polygeny, phrenology, and craniometry. But its early twentieth-century manifestation, in the work of those considered the finest scientists of the time—primarily eugenicists and geneticists—marked an important change. Whereas nineteenth-century race concepts were rooted in theories of racial distinctiveness based on measurable and observable physical traits such as cranial capacity and skin color, in the early decades of the twentieth century the biological sciences conceived of race as a reflection of unseen differences attributed to the then recently discovered factors of heredity, also known as genes. If polygeny, social Darwinism, and craniometry were the scientific backbones of a nineteenth-century understanding of race, then in the twentieth century eugenics and genetics played that same role, providing the formative language of modern racism. Hence, beliefs about racial differences became rooted primarily in biology rather than in social or economic ideologies. Over the twentieth century, the race concept had various incarnations in biology. It was modified and abandoned, embraced and repudiated by scientists. Yet it survives into the twenty-first century, persisting largely as a biological concept in both science and society. This book shows how scientists, even with the best intentions of modernizing or modifying the concept to keep with the scientific practices of the time, wound up reinforcing it and helping to ensure its survival.

Race has been called man's most dangerous myth, a superstition, and, more recently, a social construction.[1] Race concepts are rooted in the belief that the people of the world can be organized into biologically distinct groups, each with their own discrete physical, social, and intellectual characteristics. Changes to and variations in race concepts are themselves products of a range of variables, including time, place, geography, politics, science, and economics. As much as scientists once thought that race was a reflection of physical or biological differences, today social scientists, with help from colleagues in the natural sciences, have shown that the once seemingly objective race concept is in fact historically contingent and has had an unmistakable impact on the American story. Two interwo-ven histories—the introduction of and consequent use of the term "race" in the study and explication of human difference and the general use of the race concept—inform the evolution of the race concept in twentieth-century biological thought.

The historian Bruce Dain reminds us, "Race itself was a monster if ever Americans conceived one, but a monster hidden in their minds, not, as many of them came to think in the reality of a nature behind their appearances." And, as Dain is quick to point out, "that reality was obscure, shifting, and complex."[2] But one constant in that reality is that since the late eighteenth century science has played a critical role in the formula-tion of racial views in the United States, and racists and racial theorists have often turned to science to both justify their beliefs and to provide a scientific vocabulary for explaining human difference. In the twenti-eth century, it was primarily the discipline of genetics from which racial scientists freely exploited both language and prestige. This legacy can be explained largely by the history of genetics itself, which at its founding was inseparable from eugenic theories that were mired in examining heredi-tary traits both within and between human races.[3] The fields of genetics and eugenics would begin to diverge as early as the second decade of the twentieth century as geneticists in the United States sought to develop a more rigorous and less politically intent field. But despite this growing split between the two disciplines, the imprint of eugenical thinking on genetics remained strong, as did the field's reliance upon genetics. Even today, the typological thought characteristic of eugenicists at the turn of the twentieth century—that is, the way eugenicists correlated both skin color and nationality with a wide array of physical, behavioral, and intel-lectual traits—continues to be present in beliefs about human difference.

Although a genetic approach was novel to racial scientific thought in the early twentieth century, race thinking about human difference in both science and society was definitely not. The roots of race thinking had been growing in Western thought for centuries. To be sure, American ideas about race difference have been constructed in a variety of ways from numerous corners of social and scientific life, including legal, anthropo-logical, cultural, and sociological conceptions of racial difference. There are, in fact, many race concepts. So when this book refers to *the* race concept, as it often does simply for the purpose of literary parsimony, I recognize that there are others and that the concept described in this book has existed on shifting terrain, even within the nomenclature of the biological sciences. Ultimately, as this book argues, in the twenty-first century, understanding the way race was constructed within the biological sciences, particularly within genetics and evolutionary biology, is essential to understanding its broader meanings.

In many ways, this book documents the process of "racecraft," a term recently coined by Karen Fields and Barbara Fields in their eponymously named book—*Racecraft*—meant to convey the "mental terrain" and "pervasive belief" from where racism and our stubborn belief in race emanate.[4] In other words, racecraft reflects both how these ideas are sewn into our individual and collective identities and how deeply embedded in those identities are the self-reflexive assumptions that these ideas are true. Racecraft is a way of seeing, understanding, and reflecting upon our world, even when there is no rational basis for a certain worldview. The history of the race concept in American scientific thought reflects just this: the persistence of long-standing social conceptions of the meaning of difference in the thinking, theorizing, and actions of America's scientific minds. Fields and Fields's description of racecraft implicitly recognizes its permeation into scientific thought, which they explain by stating, "The term highlights the ability of pre- or non-scientific modes of thought to highjack the minds of the scientifically literate."[5] Both the eugenicists and racists who sought to utilize the race concept to buttress a discriminatory status quo *and* the liberal scientists who fought to modernize the concept were equally involved in the perpetuation of racecraft.

Histories of the race concept in American scientific thought have generally told the story of two conflicting and competing ideologies seeking to define the meaning of race within the biological sciences. On one side of this morality tale are racists, working both in and outside of scientific fields to formulate ideas about the meanings of human diversity and to propagate them under the scientific guise of racial difference. While not necessarily self-avowed racists, their agenda and actions have supported white supremacy. From Thomas Jefferson's musings on the subject in the late eighteenth century, in which he theorized that the difference between the races "is fixed in nature" and hypothesized that blacks were "originally

a distinct race," to Samuel Morton and the American School of Anthropology's nineteenth-century theories about a racial hierarchy of intelligence and of separately created races (the theory of polygeny), to eugenics
and the racialized theories of IQ over the course of the twentieth century,
racists have sought to utilize science to further their causes.[6]

On the other side of this divide have been liberal-minded scientists and
their allies who have battled the forces of racism through their scientific
work and popular writings. Theirs is a story of the rise and fall of racial
science and of the race concept itself. At the outset of the twentieth century scientific minds like W. E. B. Du Bois and Franz Boas showed that
the race concept was a social construction by illustrating how race was
a much more fluid and complex phenomenon than had previously been
thought, and that culture and economic circumstances played a more
significant role in creating the disparities between racial groups that had
been attributed to biological differences. At midcentury, anthropologists
like Ashley Montagu and sociologists like Gunnar Myrdal fought against
the race concept in their work. In 1950, the United Nations Educational,
Scientific and Cultural Organization (UNESCO) issued its first statement
on race, proclaiming, "For all practical social purposes 'race' is not so
much a biological phenomenon as a social myth."[7] These scientists battled
the racist and eugenic forces in scientific practice to push racial science to
the margins and show that it was a social construction. In other words, a
biological understanding of race has been constrained by the social context in which racial research has taken place.

This idea that there was a struggle between two fairly well-defined
groups of scientists, that racial science rose pre–World War I and waned
post–World War II, and that in this same time line race shifted from a
concept rooted in typology to one rooted in population genetics does
not hold up upon closer examination. This history, it turns out, is not so
simple and not so hopeful. The notion that the race concept and racial
science have somehow withered, or that the concept is being resurrected
by genomics and the work of the Human Genome Project, is rooted in
the post–World War II era liberal hope that by showing race to be a
social construction, the seemingly intractable problem of racism could
be overcome. The premise of a rise and fall is central to what the sociologist of science Jenny Reardon calls "the canonical narrative of the history

of race and science." It is a "dominant narrative," as she calls it, one that "truncates history."[8]

In his book reimagining the John F. Kennedy assassination, *11/22/63*, Stephen King describes history as "obdurate"—a nearly immovable force that itself fights change. The same could be said of historiography, which is also obdurate. It changes, in many ways, more slowly than the history from which it seeks to extract truths and meanings. By truncating our understanding of the evolution of the race concept, the "canonical narrative" hides a richer and much more disturbing past—one that roots a modern race concept in eugenical thought, one that examines how the race concept in biology survived many challenges (from both within science and without) and was an animating force in science throughout the twentieth century and into the twenty-first, and one that considers how even those thought to be antiracist scientists helped preserve a concept they thought they were contesting.

Scholars have begun to contest this canonical narrative. Works by Jenny Reardon on the history of the Human Genome Diversity Project, Gregory Dorr on the relationship between eugenics and segregation in Virginia, and Lee Baker on the role that anthropologists have played in the reformulation of concepts of race all reframe how we think about this history.[9] In this book I build on these works and others, arguing that the biological race concept, as we understand it today, originated with eugenic theories of difference and was re-created and integrated into modern biological thought by population geneticists and evolutionary biologists in the 1930s and 1940s during the evolutionary synthesis in biology (the union of population genetics, experimental genetics, and natural history that reshaped modern biology).

While important changes in the biological approach to race did occur as early as the 1930s, particularly as an increasing number of geneticists, anthropologists, and social scientists began moving away from typological and eugenic descriptions of human difference to view races through the lens of population genetics and evolutionary biology, the shift away from typology was not as complete and was much more complicated than the canonical narrative suggests. Contrary to so much of the literature on the race concept, the field's shift on race was not simply the liberal triumph of science over ignorance. Instead, it was first a struggle to find meaning for the concept within taxonomic nomenclature and the evolutionary synthesis, and, second, a struggle to find alternative ways to explain human genetic diversity. And it was in this contradictory space that a growing group of scientists found themselves as they struggled to both find meaning for a race concept in science and fight against racial science and racism more generally.

Many, in fact, came to reject a eugenic and typological notion of fixed genetic differences between so-called racial groups and instead understood human races as dynamic populations distinguished by variations of the frequency of genes between them. By rooting the meaning of race in genetic variation it became more difficult (though still possible) to root race in eugenic conceptions of difference and to argue that one race or another had particular traits specifically associated with it, or that one individual was typical of a race. Furthermore, the four or five racial groups identified by eighteenth- and nineteenth-century scientists now varied depending upon the genes and traits examined by geneticists. Theodosius Dobzhansky, the evolutionary biologist whose work between the 1930s and 1970s had a tremendous influence on the way that scientists thought about race, concluded that the number of human races was variable depending upon what traits were being examined. In fact, Dobzhansky believed the race concept in the context of population genetics and evolutionary biology was simply a tool for making genetic "diversity intelligible and manageable" in scientific study.[10] In other words, while human differences are real, the way we choose to organize those differences is a methodological decision and not one that reflects an underlying evolutionary hierarchy.

This new approach was brought about by novel findings in genetics that demonstrated that genetic variation was much more common within species than once thought, and by the development of the evolutionary synthesis, which rejected eugenic notions of difference between and among species. Changes in the race concept were also influenced by a growing cadre of scientists who were generally more liberal on matters of race than their predecessors had been, as well as by a gradual liberalization on matters of race in post–World War II America. Indeed, as this book documents, this was a two-way street—the scientists involved in conceptualizing a race concept in biology were as much a product of the scientific culture in which they were trained as they were a part of the social milieu in which they lived their lives. Unfortunately, what was believed to be the methodological utility to evolutionary biologists and population geneticists of this new race concept would help reinforce confusion about the term, even within the field, and it would quickly be exploited and manipulated by racists from both within and outside the field. By the 1960s, Dobzhansky, whose work helped re-create race in the framework of population genetics and evolutionary biology, came to the conclusion that despite race's utility as a tool for classification and system-atization—"devices used to make diversity intelligible and manageable"—that investigation into human diversity had "floundered in confusion and misunderstanding." He also came to believe that the scientific and social meanings of race were inseparable, and that "the problem that now faces the science of man is how to devise better methods for further observa-tions that will give more meaningful results."[11]

Racial science did not simply end with the decline of the eugenics movement in the 1930s and 1940s, which was brought down by advances in scientific thinking that recognized the fallacy of the eugenic proposi-tion and by a worldwide reaction to Nazi eugenical horrors. Nor did it recede in the wake of statements on race and racism by UNESCO in the early 1950s—statements that were critical of the race concept and helped to shape thinking in this area among both natural and social scientists. Nor did the psychologist Kenneth Clark's studies illustrating the effects of segregation and white supremacy on African Americans, studies that fig-ured prominently in the Supreme Court's 1954 landmark *Brown v. Board of Education* ruling, bring an end to racial science.[12] Instead, racial sci-ence and the race concept have survived many intellectual and political challenges. The historian William Stanton once said of the race concept that "man was being fitted into a system of immutable law."[13] When biolo-gists at midcentury reaffirmed the race concept in the context of modern genetics, they were, intentionally or not, preserving racist ideas in science for both scientific and extrascientific purposes.

A history of the race concept in biology would be incomplete with-out understanding the role that eugenics played in its development. To a reader well versed in the eugenic literature, many of the characters and issues raised in chapters 1 and 2 of this book will seem quite familiar.

While it is true that a major objective of the eugenics movement was to keep the "unfit" from reproducing, it is also true that the movement and its architects helped develop a new language of difference and, therefore, of race in the twentieth century. This facet of eugenics has been largely overlooked in the historical literature. It would not be accurate, however, to suggest that eugenicists simply reflected the deeper racial anxieties and animosities of the nation in the early decades of the Jim Crow era. This book shows how, instead, eugenicists actually helped shape the way in which those animosities were incorporated into the scientific lexicon of the times.

A rereading of eugenic-era primary source materials reveals that incorporation and revises that history by showing how beginning in the late nineteenth and early twentieth centuries eugenics focused intently on the black-white divide in American society and sought to explain that divide in eugenic terms. It also shows how the attention of eugenicists to what they believed to be fundamental differences between whites and blacks provided a foundation for rethinking the race concept during the first three decades of the twentieth century. Eugenicists, as this book shows, devoted considerable resources to the study of black-white differences from the beginning of the movement in the late nineteenth century. Yet historians have painted eugenics as largely incidental to the formulation of ideas of race in science, focusing instead on the history of eugenic institutions, on the relationship between eugenics and emerging conceptions of ethnicity among immigrant white groups, and on the impact of eugenic policies (in particular sterilization programs and immigration restrictions). These approaches overlook the links between eugenic thought and ideologies of race and racism and their impact on African American history.

Attention to the impact of racial science on African Americans during the twentieth century has focused heavily on two histories: the Tuskegee Experiments (Study of Untreated Syphilis in the Negro Male) and IQ studies. By telling these stories separately, historians have often missed a bigger picture—a narrative exploring how, beginning with eugenics, biology has been used to buttress and rationalize American's changing view of African Americans, and that thinking in the natural sciences has influenced the continued evolution of racist ideology in the United States.

It is also significant that before 1924 explanations of racial difference by eugenicists and geneticists, according to most historians, were focused primarily on differences among what we would now consider white ethnic groups. Eugenics was not just about preserving whiteness from ethnics but was also about the construction of scientifically justified color differences. In the wake of the Johnson-Reed Act of 1924 (also known as the Immigration Act of 1924), which severely restricted immigration into the United States (legislation urged and supported by eugenicists), the focus of many eugenicists and other racial scientists shifted toward examining black-white racial differences.[14] An examination of the discussions and debates taking place at this time among and between eugenicists, geneticists, evolutionary and population biologists, and anthropologists reveals how academic thinking helped to formulate the science behind ideologies of race and racism.[15]

The race concept has had a marked impact on the practice of science and on the social understandings of human difference from eugenics to genomics. By examining the history of the biological race concept during the twentieth century, historians have borne witness to the ways in which the biological sciences have helped to shape thinking about human difference. The historian Charles Rosenberg reminds us that "science has lent American social thought a vocabulary and supply of images."[16] This book describes the role that scientific thought, particularly genetics, played in developing a language and methods used to measure the meaning of human difference in the form of race. The philosopher Kwame Anthony Appiah explains that the diffusion of scientific ideas and concepts into the general population took place by a process he calls "semantic deference"; "that is, with the increasing prestige of science, people became used to using words whose exact meanings they did not need to know, because their exact meanings were left to the relevant experts."[17] Through these processes, during the twentieth century biology and genetics became an arbiter of the meaning of the race concept.

An examination of the history of the biological race concept reveals that race is not what most people think it is. Ultimately, as this book illustrates, race is neither a static biological certainty nor a reflection of our genes. Instead, race is a historical and cultural phenomenon—an analysis of human biological difference mediated by the politics, culture, and economics of a given historical moment and by the individual or society in that moment. For America, the corrupting power of racial thought remains embedded in its social structures. We see it in disparities in health, in housing, and in employment and social opportunities.[18] This work does not claim to expose the nature of these disparities or how to mitigate them; this work focuses instead on the ideas behind the race concept. But by examining the historical and intellectual bases for the race concept, we can, it is hoped, begin to understand its origin and develop news ways of thinking about the meanings of human diversity.

This book shows how the biological race concept came to be what it is today and why, because of this history, race continues generating controversy as a classificatory tool.

# 1

# A EUGENIC FOUNDATION

"There is, unquestionably, a larger popular interest in races and racial traits now than ever before," claimed Charles Davenport in 1921. A biologist by training, Davenport was the leader of the American eugenics movement during the first three decades of the twentieth century and wrote and lectured widely on the subject. "For some people race seems to be equivalent to European country of origin," declared Davenport, echoing in his statement what at the time was a race concept that conflated skin color with nationality. Those, for example, of Italian, Polish, and German descent were popularly and often scientifically believed to belong to Italian, Polish, and German races. Davenport also described how the "Immigration Bureau recognizes 'races or peoples' such as Hebrew" and how "the U.S. census has classified the population by 'mother tongue,'" which, he asserted, was "biologically slightly more significant than country." In recognition of these seeming contradictions, Davenport worried that "ultimately little attention is paid to the question: what is a race and how do you define any particular race." "There would seem to be a need for a reconsideration of the idea of race and the definition of particular races," Davenport asserted, concluding that "men of science are looked to for such clearer ideas and definitions."[1] It would be from the eugenics movement, led in Britain by Francis Galton in the late nineteenth century and in the United States by Davenport during the first three decades of the twentieth century, that "men of science" would address the challenge of defining this problematic concept.

While ideas about biological distinctiveness were a part of the racial lexicon since the early nineteenth century, eugenics offered a scientific explanation for racial difference. Eugenicists correlated certain negative and deviant social behaviors—including criminality, insanity, and feeble-mindedness (a term that captured any number of mental disorders)—with particular ethnic and racial populations, and claimed these behaviors to be inherited via the gene.[2] In the first three decades of the twentieth century eugenicists and their supporters applied such ideas about racial difference to immigration, reproductive, and racial policies. The geneticization of race—the idea that racial differences can be understood as genetic distinctions in appearance and complex social behaviors between so-called racial groups—came about in the wake of the eugenics movement.

Eugenicists differed on how best to repair what they saw as a dysgenic society filled with what they believed to be genetically unfit groups, including most prominently immigrants from eastern and central Europe, who were outreproducing Americans of northern European ancestry. Whereas positive eugenicists sought to increase breeding among the American social elite, negative eugenicists, in contrast, discouraged breeding among the lower classes.[3] During the heyday of the eugenics era in America, popular culture and policy enactments were dominated by the theories of negative eugenicists. Positive eugenics was the terrain primarily of social-ist intellectuals who believed that eugenics would facilitate the emergence of a socialist utopia in the United States.[4]

Sterilization laws across America were inspired by negative eugenic sentiment, and in 1907 the state of Indiana established the country's first sterilization law. By the early 1930s over twenty-nine other states would pass similar laws, leading at that time to the sterilization of approximately 30,000 so-called feebleminded Americans. That figure would rise to total more than 63,000 sterilizations by the 1960s.[5] Criminals and those accused or convicted of sexual offenses were the primary concern of these eugenic enactments. Advocates of criminal sterilization wrote, "Criminals should be studied for evidence of dysgenic traits that are germinal in nature. Where found in serious degree parole should not be granted without sterilization."[6]

In the first three decades of the twentieth century eugenicists and many geneticists promoted the idea that mental and physical traits differed hereditarily by race. They also claimed that race crosses were harmful.[7] Well-respected geneticists wrote openly that "miscegenation can only lead to unhappiness under present social conditions and must, we believe, under any social conditions be biologically wrong."[8] In the late 1920s Davenport wrote, "We are driven to the conclusion that there is a constitutional, hereditary, genetical basis for the difference between the two races [whites and blacks] in mental tests. We have to conclude that there are racial differences in mental capacity."[9] In their influential text *Applied Eugenics*, Paul Popenoe and Roswell Hill Johnson, who endorsed segregation as a "social adaptation," wrote that "the Negro race differs greatly from the white race, mentally as well as physically, and that in many respects it may be said to be inferior when tested by the requirements of modern civilization and progress." Moreover, they suggested that "Negroes, both children and adults, have been found markedly inferior to white in vital capacity. . . . Differences in temperament and emotional reaction also exist, and may be more important than the purely intellectual differences."[10] Through eugenics, genetics gave race and racism an unalterable permanence; neither education, nor change in environment or climate, nor the eradication of racism itself could alter the fate of African Americans or those labeled as belonging to nonwhite races.

There were, to be sure, even in the eyes of the most racist thinkers, exceptions to black genetic inferiority. But eugenicists and other scientific racists explained these "aberrations" by noting that genetic material from white ancestry set them apart. W. E. B. Du Bois's success was, for example, attributed to the blood he inherited from his white ancestors.[11] In this context, it is not hard to see how eugenics provided a modern scientific language, rooted in the burgeoning field of genetics, that both proffered and buttressed contemporary racial theories. The legacy of eugenics therefore is not simply about sterilization laws, anti-immigration statutes, or its impact on Nazi racial theory. Those events, important issues in their own right, have been explored by numerous historians.[12] Ultimately, eugenics in an immigrant and ethnic context was about social control. But in a black versus white context eugenics was also about defining (with the latest scientific theory and jargon) the nature of the social and biological differences believed to be reflected by skin color.

## FRANCIS GALTON AND THE FOUNDATIONS OF THE
## BLACK-WHITE BIOLOGICAL DIVIDE

From the works of Francis Galton in England from the 1860s to the 1900s, to the corpus of writings of Charles Davenport in the United States from the 1900s to 1930s, eugenicists showed a keen and consistent interest in using their ideas and methodologies to understand racial differences between blacks and whites on both sides of the Atlantic, and thus played a fundamental role in the construction of American concepts of race and racism in the early decades of the twentieth century. The focus of eugenic ideas on hereditary differences between whites and blacks has been part of the eugenic literature since the earliest days of the movement.[13]

Galton, founder of eugenics, published his first essay in the field, "Hereditary Talent and Character," in *Macmillan's Magazine* in 1865. Interestingly, when Galton wrote the *Macmillan's* article he had not yet even coined the term "eugenics."[14] The article focused primarily on his early ideas about how human traits passed between generations, on which Galton wrote, "Our bodies, minds and capabilities of development have been derived from them [our forefathers]." In the late nineteenth century the secrets of heredity had not yet been revealed except for the relatively obscure work of the monk-scientist Gregor Mendel, whose laws of heredity would not be rediscovered until the first years of the twentieth century.

As the scion of a prominent family that included his maternal grandfather, the great physician, inventor, and naturalist Erasmus Darwin, and his cousin, the celebrated naturalist and architect of evolutionary theory, Charles Darwin, Galton believed biological heritage to be of profound importance in his life. Through eugenics, he theorized that heredity exerted a singular influence on all the social characteristics of humankind. With such bloodlines, it seems no coincidence that Galton's "inquiries into hereditary genius . . . show the pressing necessity of obtaining a multitude of exact measurements relating to every measurable faculty of body or mind, for two generations at least, on which to theorize."[15] With this information Galton hoped to improve the world through selective breeding. Karl Pearson, Galton's star pupil, a famed eugenicist himself, and founder of the Galton Laboratory at the University of London, summarized what he believed to be Galton's vision of a eugenic world: "Democracy—moral and intellectual progress—is impossible while man is burdened with the heritage of his past history. It has bound mankind to a few great leaders; it has produced a mass of servile intelligences; and only man's insight—man breeding man as his domesticated animal—can free mankind."[16]

From a very young age, according to biographer Raymond Fancher, Galton's parents, "who collected and saved documents as evidence . . . regarding Francis's precocity in their diaries," had high intellectual expectations for young Francis, and he was "cast firmly in the role of family academic from the time of his first glimmerings of scholarly aptitude."[17] The young Galton's intellectual feats were self-recorded in a letter to his older sister when he was four: "I am four years old and can read any English book. I can say all the Latin Substantives and Adjectives. . . . I can cast up any Sum in addition and can multiply by 2, 3, 4, 5, 6 7, 8, 9, 10."[18] Tutored by his older sister until he was five, Galton was educated at prep schools until sixteen, when he became an apprentice to William Bowman, a young surgeon.[19] Galton began his academic medical training in 1839 at King's College, London, but at the urging of his cousin Charles Darwin, Galton left King's in 1840 to pursue mathematics and classics at Cambridge.[20] After three years at Cambridge, having failed to qualify for honors, Galton left for London to continue his medical studies. But that, too, Galton left incomplete, never becoming licensed, and his departure from medical training brought his "formal" academic career to an end.[21]

Galton, according to Fancher, spent much of the next six years living "the life of the idle rich," going on "a carousing tour of Egypt and the Middle East," where he practiced his hunting and shooting skills.[22] His transformation into a man of science, according to one apocryphal account, suggests that his life began to take shape only after a London phrenologist suggested he could build character and strength through military service in the British colonies. With money from his family inheritance, Galton set out to explore and map regions of what is now Namibia in southwestern Africa. For that effort, in 1854, Galton was awarded a gold medal from the Royal Geographic Society.[23]

Galton's time on the African continent, according to historian Daniel Kevles, "exemplified the joining of foreign adventure with scientific study," and certainly shaped his attitudes toward Africa, Africans, and the descendants of that continent.[24] In his memoir, *Memory of My Life*,

Galton's earliest connections between race and breeding are unequivocal. Of his encounter with the Damara people in southern Africa, Galton wrote that they "were for the most part thieving and murderous, dirty, and of a low type; but their chiefs were more or less highly bred."[25] Galton's assessment of African intelligence was also severe. Of his visits with the Herero people in southern Africa Galton wrote, "Whatever they may possess in their language, they certainly use no numeral greater than three. When they wish to express four, they take to their fingers. . . . They puzzle very much after five, because no spare hand remains to grasp and secure the fingers that are required for the 'units.'" But Galton, like many of his contemporaries, expressed his views about race through science, and as his eugenical ideas took shape, his social prejudices became scientific ones. Galton's personal observations about Africans would, in the coming years, evolve into more careful scientific thinking on the subject. His racism would also be obvious in many of his writings, including his two seminal works, *Hereditary Talent and Character* and *Hereditary Genius*.[26]

Galton accepted the idea that physical and social traits were associated, despite his not having any evidence to support this point.[27] Galton extended this idea to human racial types, writing in his 1865 essay on heredity in *Macmillan's*, "Mongolians, Jews, Negroes, Gipsies, and America Indians severally propagate their kinds; and each kind differs in character and intellect, as well as in colour and shape from the other four."[28] In the *Macmillan's* article Galton paid careful attention to blacks, suggesting, for example, "the Negro has strong impulsive passions, and neither patience, reticence nor dignity." Galton also maintained that "the Negro" is "warmhearted, loving towards his master's children and idolized by the children in return. He is eminently gregarious, for he is always jabbering, quarrelling, tom-tom-ing and dancing. He is remarkably domestic, and is endowed with such constitutional vigour, and is so prolific that his race is irrepressible."[29] It would be easy to dismiss these racist musings as simply a reflection of Victorian views on race, which prevailed among both British and Americans at this time. Some have even argued that Galton's writings on Africans and blacks "occupied only a minuscule fraction of his writings on human heredity." But Galton's attention to race, specifically as it concerned blacks, was, in fact, significantly developed and nuanced, and as the founder of a movement of which the "betterment of the race" was a core principle, Galton's writings on this matter would have been read widely with considerable effect on the shaping of eugenics.[30]

In mid to late nineteenth-century England, Galton probably had few direct interactions with persons of African ancestry. Galton's experiences with blacks came from his earlier travels in Africa and what we must assume to be his limited contact with the few blacks living in London at the time. The black population of London shrank during the latter part of the eighteenth and into the first several decades of the nineteenth century following the ending of the slave trade in 1807 and the subsequent absorption of Africans and their descendants into English society.[31] Great Britain itself was certainly not immune to the struggles around issues of slavery that had plunged the United States into civil warfare during the time Galton began to spell out his eugenical theories in his writings. The impact of both domestic and international racial conflict would be significant on Galton's thinking in this area.[32]

On the effect of slavery on Britain's views of black Africans, Sir Richard Burton, the renowned nineteenth-century British adventurer and explorer, declared, "Before the Wilberforcean age, he was simply a Negro. That trade which founded in Liverpool, and which poured five million pounds of sterling into the national pocket, marked him to the one class a Man and a Brother, to the other a Nigger."[33] Views of blacks would change little after the emancipation of African slaves in Great Britain in 1807. In 1849, an anonymous author published an essay in *Fraser's Magazine* titled "The Nigger Question." The writer believed the typical black to be the lowest of the human species and wrote of their future, "Decidedly you will have to be servants to those that are born wiser than you, that are born lords of you; servants to the whites, if they are (as what mortal can doubt they are?) born wiser than you."[34] In the succeeding issue of *Fraser's* the philosopher John Stuart Mill offered a challenge to racist theories, writing that racial differences between blacks and whites were produced by circumstance, not nurture. Just a year earlier, Mill had written about the nature of the racist argument, stating, "Of all vulgar modes of escaping from the consideration of the social and moral influences on the human mind, the most vulgar is that of attributing the diversities of human conduct and character to inherent original natural differences."[35]

Galton, who joined the Ethnological Society of London in the early 1860s, was likely influenced by debates at the society on the origin of human races.[36] At that time, its members were arguing about the scientific legitimacy of a polygenic view of mankind—whether or not blacks and other racial groups were actually a distinct species. In 1863, for example, the president of the society, in an address called "The Negro's Place in Nature," argued in favor of a polygenic view, concluding that blacks were a distinct race, were intellectually inferior to whites, and that European civilization was "not suited to the Negro's requirements or character."[37] Polygenic theories were popular on both sides of the Atlantic at this time, driven largely by the efforts of Samuel Morton and the American School of Anthropology.

In the closing decades of the nineteenth century, Britain's colonial involvement in Africa grew deeper, justified in large part by such anthropological and biological ideas. Galton's prejudices against Africa were on full display as he weighed in on the colonization debate in an essay published in the *Edinburgh Review* in 1878. He hoped for Africa that "men of other races than the negro, such as the Chinese coolie," would "emigrate, and, by occupying parts of the continent . . . introduce a civilisation superior to that which at present exists."[38] On the interplay between politics and race on the African continent, Galton suggested, "The recent attempts by many European nations to utilise Africa for their own purposes gives immediate and practical interest to inquiries that bear on the transplantation of races. They compel us to face the question as to what races should be politically aided to become hereafter the chief occupants of that continent."[39] Ultimately, in Galton's mind, the inferiority of Africans predetermined that outcome as he expected that "it may prove that the Negroes, one and all, will fail as completely under the new conditions as they have failed under the old ones, to submit to the needs of a superior civilisation to their own; in this case their races, numerous and prolific as they are, will in course of time be supplanted and replaced by their betters."[40] It is worth considering whether the British sociologist Michael Banton was correct in concluding that "the imperialist philosophy could ever have taken such a hold upon the nation's mind had it not been for the development of certain anthropological and biological doctrines."[41]

Galton was a prolific author, and his writings helped him popular-ize the study of the betterment of racial groups through eugenics. In his first book-length work, *Hereditary Genius,* published in 1869, Galton's primary concern was "whether or no genius be hereditary," and claimed "to be the first to treat the subject in a statistical manner."[42] The book received mixed reviews in both the scientific and popular press. Writ-ing in *Nature,* Alfred Russel Wallace, who, along with Charles Darwin, is jointly credited with uncovering the mechanisms of evolution, wrote that those "who read it without the care and attention it requires and deserves, will admit that it is ingenious, but declare that the question is incapable of proof." The *London Daily News* embraced the book, writing that "Galton undertakes to show, and to a large extent undoubtedly suc-ceeds in showing, that genius is equally transmissible, and that ability goes by descent." Both the *London Times* and the *Saturday Review* were unflattering in their reviews. The *Times* reviewer wrote, "Galton is a little too anxious to array all things in the wedding garment of his theory, and will scarcely allow them a stitch of other clothing." The *Saturday Review*'s assessment suggested that Galton "bestowed immense pains upon the empirical proof of a thesis which from its intrinsic nature can never be proved empirically."[43] Galton's attraction to studying questions of heredity was influenced, in large part, by the work of his cousin Charles Darwin. While Galton had accepted Darwinian evolutionary theory, he searched for alternative methods by which evolution occurred. Whereas Darwin theorized that evolution was gradual and continuous, Galton believed it to be abrupt and discontinuous.[44] Some have even speculated that mod-ern theories of heredity, including eugenics, were launched to either chal-lenge or complete Darwin's theory of evolution.[45]

In a chapter in *Hereditary Genius* with the title "The Comparative Worth of Different Races," Galton proposed that "every long-established race has necessarily its peculiar fitness for the conditions under which it has lived."[46] It is therefore at the level of racial groups that intelligence will have its most significant impact. "Among animals as intelligent as man, the most social race is sure to prevail, other qualities being equal," sug-gested Galton.[47] In measuring the "worth of races," Galton made "use of the law of deviation from an average," a law that was both the centerpiece of eugenics and also represented Galton's lasting contribution to the field of mathematical statistics. Only the "Australian type" made out worse than the "African negro," the former considered by Galton to be one grade below the residents of the African continent and their descendants.[48]

Galton introduces the reader to "comparative racial worth" by comparing "the negro race with the Anglo-Saxon, with respect to those qualities alone which are capable of producing judges, statesmen, commanders, men of literature and science, poets, artists, and divines."[49] While Galton acknowledged "the negro race is by no means wholly deficient in men . . . considerably raised above the average of whites" (citing as proof, for example, Toussaint Louverture, leader of the Haitian Revolution), he concluded that "the average intellectual standard of the negro race is some two grades below our own." Galton also cited a statistic suggesting "the number of negroes of those whom we should call half-witted is very large." Recalling his own visit to Africa, Galton remarked that "the mistakes the negroes made in their own matters, were so childish, stupid, and simpleton-like, as frequently to make me ashamed of my own species."[50] Galton's rhetoric on this subject offers no statistical "proof" or data for these assumptions, yet reading his observations about the "lowest" or "highest" races, one is struck by the language of statistical certainty permeating his writing.

Fourteen years later, with the publication of *Inquiries into Human Faculty and Its Development*, Galton expanded his hereditary focus to also include physical, social, and mental traits. Although he had been writing about the idea of eugenics for the almost two decades since the publication of his *MacMillan's* essay, in *Inquiries* Galton first offered a definition for his burgeoning field of investigation. Early in the text of the book, he introduced eugenics simply as "the cultivation of race."[51] Just below the definition, however, in an extended footnote, Galton notes the Greek origins of the word and shares with the reader his enduring vision for the field, writing that "we greatly want a brief word to express the science of improving stock, which is by no means confined to questions of judicious mating, but which, especially in the case of man, takes cognisance of all influences that tend in however remote a degree to give to the more suitable races or strains of blood a better chance of prevailing speedily over the less suitable than they otherwise would have had. The word *eugenics* would sufficiently express the idea."[52]

In *Inquiries* Galton wrote with conviction about what he believed to be the "fact" that "the very foundation and outcome of the human mind is dependent upon race, and that the qualities of races vary, and therefore that humanity taken as a whole is not fixed by variable, compels us to reconsider what may be the true place and function of man in the order of the world."[53] To satisfy that conviction, Galton used the pages of *Inquiries* to begin to develop a program (that is, of course, eugenics) by which to improve different races of humanity. "My general object has been to take note of the varied hereditary faculties of different men, and of the great differences in different families and races," wrote Galton. He hoped to use this information to investigate the "practicability of supplanting inefficient human stock by better strains, and to consider whether it might not be our duty to do so by such efforts as may be reasonable."[54] For Galton, his supporters, and many of his peers, eugenics was a utopian method through which to breed better humans. A reviewer in *Science* noted and endorsed the utopian nature of Galton's proposed method, writing, "If we want human stock to grow better through voluntary effort, we must undertake to study and improve pre-natal and ancestral influences yet more than we try to better the influences of education."[55]

Ultimately, however, the language and content of eugenics could be about using the scientific method for advancing and preserving white supremacy. Where *Hereditary Genius* established a eugenic hierarchy of races, *Inquiries* offered explanations for why this was so and why it would stay this way. Galton believed that "so long as the race remains radically the same, the stringent selection of the best specimens to rear and breed from, can never lead to any permanent result." Galton likened the struggle to improve a low race through noneugenic means to "the labour of Sisyphus in rolling his stone uphill; let the effort be relaxed for a moment, and the stone will roll back." Instead, the only way to improve a low race was by allowing only "the few best specimens of that race . . . to become parents, and not many of their descendants can be allowed to live."[56] This chilling passage, the spirit of which was adapted in the 1930s by the Nazis as part of their Final Solution, is the eugenic extreme. Similarly, this new scientific view of race was adopted and perpetuated by Galton's acolytes, by racists, and by white supremacists on both sides of the Atlantic for more than a century.

It is fairly easy to show through documentary evidence how Galton theorized about racial differences and, more specifically, about what he believed were black and white racial differences. There is more than sufficient evidence illustrating this point. However, even more important than his eugenical thinking about blacks is to recognize that Galton and his eugenic followers fundamentally changed the meaning and study of race and racism. If the historian Ruth Schwartz Cowan is right, and "Galton changed the study of heredity by changing the meaning of the word 'heredity,'" then we must also consider how Galton and his eugenicist disciples also changed the study of race by changing the meaning of the word "race."[57] Schwartz Cowan argues that "heredity," prior to Galton and eugenics, was "a word which had long been poorly, if not vaguely, defined."[58] Prior to Galton's writings, words like "inheritance" and "hereditary" were used to generally describe intergenerational legacies. But according to Cowan, "the passage from 'inheritance' to 'heredity' meant passing from an extremely flexible definition, one which was so vague as to be of little scientific value, to an extremely concrete definition, one which may have been overly rigid but which was nonetheless quantifiable, explorable and researchable."[59]

Galton's writings exerted a similar influence on race, and, as with his impact on the study of heredity, he and a generation of eugenicists redefined the term and its study. If eugenics was, at its core, about quantifying heredity, then through the lens of eugenics, race became the most important quantifiable human trait to study in a hereditarian context.[60] Galton was not alone in believing that the "survivorship of the fittest" would occur at the level of race. The term "race" itself had long been the subject of debate among the nineteenth century's anthropologists, biologists, and philosophers. Whereas nineteenth-century anthropology failed to offer a lasting scientific vision of racial difference through the theory of polygeny—polygeny posited a divine hierarchy of separate creations, a theory that contradicted the Bible—eugenicists found a way to quantify and reify racial attitudes without undermining or challenging accepted religious mores. Galton's success was in developing mathematical methods to study human diversity in the context of heredity. Galton called eugenics "the science which deals with all influences that improve the inborn qualities of a race; also with those that develop them to the utmost advantage."[61] By pairing heredity and race under the banner of eugenics, Galton was able to redefine how race was employed in a scientific context.

Nineteenth-century anthropologists, including most prominently Samuel Morton, based theories of racial distinctiveness on measurable and observable physical traits. Their approach to measuring human difference was typological, under the belief that traits like cranial capacity and skin color were correlated with specific intellectual and behavioral characteristics. Eugenicists took a different approach; they shifted to seeing and measuring race as a reflection of unseen differences they attributed to heredity, an area of study they would help to create in the final decades of the nineteenth and early decades of the twentieth centuries. This shift, from the seen to the unseen, which in today's genetic parlance would be from the phenotypic to the genotypic, was the eugenicists' most significant contribution to redefining the meaning of race. By the early twentieth century, as genetics became the field through which to study heredity, eugenicists helped to geneticize the study of human diversity.

Rooting human variation in blood or in kinship is a relatively new way to categorize humans. The idea gained strength toward the end of the Middle Ages as anti-Jewish feelings, which were rooted in an antagonism toward Jewish religious beliefs, began the long evolution into anti-Semitism, which rationalized anti-Jewish hatred as the hatred of a people. For example, Marranos, Spanish Jews who had been baptized in the Church, were still considered a threat to Christendom because they could not prove purity of blood to the Inquisition. Despite an outward acceptance of Christ, a Jew would always be a Jew.[62]

During the Enlightenment, ideas of this type took root and were more directly applied to explaining the diversity of humankind. The integration of social and cultural notions of personhood into a belief in static human communities came to fruition at this time, driven in part by the experiences with new peoples during colonial exploration, the need to rationalize the inferiority of certain peoples as slavery took hold in a protocapitalist world, and the development of a modern scientific taxonomy that provided a new type of language to assess and explain human and organismal diversity.

The progenitor of modern taxonomic classification, Swedish botanist and naturalist Carolus Linnaeus, devised his *Systema Naturae* (1735), in which "all living forms" are classified by "genus" and "species." Linnaeus defined species as "fixed and unalterable in their basic organic plan," while varieties within the species could be caused by various external factors such as climate or temperature. Linnaeus divided the human species into four groups: *Americanus, Asiaticus, Africanus, and Europeaeus.* And to these groups (he did not refer to them as races) he ascribed both physical and behavioral characteristics. Members of *Americanus* were "reddish, choleric, and erect; hair black . . . wide nostrils . . . obstinate, merry, free . . . regulated by customs." Those of *Asiaticus* were "melancholy, stiff; hair black, dark eyes . . . severe, haughty, avaricious . . . ruled by opinions." *Africanus* were "black, phlegmatic . . . hair black, frizzled . . . nose flat; lips tumid; women without shame, they lactate profusely; crafty, indolent, negligent . . . governed by caprice." Finally, those of the *Europeaeus* category were "white, sanguine, muscular . . . eyes blue, gentle . . . inventive . . . governed by laws."[63]

Linnaean taxonomy was infused with judgments of inferiority and superiority. The idea of a "Great Chain of Being"—a vision of the universe that in a hierarchical fashion ranked forms of life from the simplest to the most complex—is reflected in Linnaeus's characterization of human diversity. This Great Chain of Being "was the conception of the plan and structure of the world which, through the Middle Ages and down to the late eighteenth century," according to Arthur Lovejoy, and was accepted "without question" by "many philosophers, most men of science, and, indeed, most educated men."[64] The integration of human diversity into the Great Chain also reflected "the growing influence of a certain type of thinking, which presumed that each species had qualities of behavior or temperament that were innate." Linnaeus's own experience with different peoples was limited primarily to reports from others who had spent time in the colonies. The perceptions of these observers, according to anthropologist Audrey Smedley, "flowed into the scientific establishment and fueled its speculations."[65]

Out of this Enlightenment mix of exploration, colonization, science, and slavery emerged a modern notion of race. But this Great Chain lacked a unifying term to connote the innate sense of being (referring to both physical and social traits) in populations of peoples that Linnaeus and others beginning in the fifteenth century were trying to describe, first throughout

Europe and eventually to define and rank the peoples on the continents the Europeans colonized. Scholars are in general agreement that etymologically "race" was a latecomer to Romance and other Western languages, and that it originated in the Middle Ages as a term used primarily in domestic animal breeding to describe breeding lines or groups of animals bred for similar uses. Yet its use then, as now, remained confused and often contradictory. It could be a term simply to describe a group of people united by common characteristics, or it could be used when classifying different human groups, sometimes even in place of the term "species." It was thus not unusual to see the term describe different things: the white race, the race of Englishmen, the human race, or its use to describe family lineage.[66]

And it is from this etymological disorder that the term "race" was introduced into the natural sciences in 1749 by the French naturalist Georges-Louis Leclerc, comte de Buffon, in *Natural History*. Buffon saw clearly demarcated distinctions between the human races that, he believed, were caused by varying climates.[67] On the one hand, Buffon recognized the unity of mankind as a species, and the ability of humans to thrive while breeding between races. Buffon wrote, "The Asiatic, the European, and the Negro produce equally with the American. Nothing can be a stronger proof that they belong to the same family, than the facility with which they unite to the common stock." And Buffon also believed that while "the blood is different" between peoples, "the germ is the same." Thus, despite outward differences in physical type, and even differences in the blood, Buffon believed that the "germ" of humanity was ultimately identical. Nevertheless, Buffon's climatological theory of difference was infused with notions of European superiority. In *Natural History* he wrote, "As if, by any great revolution, man were forced to abandon those climates which he had invaded, and return to his native country, he would, in the progress of time, resume his original features, his primitive stature, and his natural color." To Buffon, this natural state of humanity was derived from the European, a people that "produced the most handsome and beautiful men" and represented the "genuine color of mankind."[68]

The impact of German scientist Johann Blumenbach's racial classifications, developed toward the end of the eighteenth century in his book *On the Natural Variety of Mankind*, continue to have a significant impact on the idea of race in modern times. Whereas Linnaeus suggested four racial types, Blumenbach offered five: Caucasian, Mongolian, Ethiopian, American, and Malay. Blumenbach's addition, noted paleontologist Stephen Jay Gould, "radically changed the geometry of human order from a geographically based model without explicit ranking to a double hierarchy of worth oddly based upon perceived beauty and fanning out in two directions from a Caucasian ideal." Even though Blumenbach "stoutly defended the mental and moral unity of all peoples," his "racial geometry" was imbued with a sense of white superiority.[69] And while racial differences were fast becoming part of the scientific vernacular, prejudice and discrimination based on skin color both preceded and complemented scientists' providing a vocabulary to racial ideology.

Galton had no way to directly measure the "unseen" differences he attributed to racial difference. However, his theory of heredity assumed them to be real reflections of the physical and social differences that he claimed he could measure. In *Inquiries into Human Faculty* Galton wrote that "we cannot but recognise the vast variety of natural faculty, useful and harmful, in members of the same race, and much more in the human family at large, all of which tend to be transmitted by inheritance."[70] Galton's methods of measuring racial difference ultimately differed little from the anthropological and typological approaches that preceded him. After all, the technologies of genetics were still a few decades away when Galton began theorizing on heredity. But what Galton did do was shift thinking about difference itself to a hereditarian worldview. While "race" remained a term with a multitude of definitions, Galton and the eugenicists succeeded in simultaneously quantifying the study of race through eugenics—Galton and his colleagues developed statistical methods to measure mental, physical, and social traits that their data showed varied by racial groups—and, paradoxically, using the term "race" in a multitude of ways that were often contradictory.

Thus, given the term's etymological and scientific history, that the inconsistent use of the term "race" is universal in the eugenic literature should not be surprising. In an 1878 article Galton wrote, "The negro may himself disappear before alien races, just as his predecessors disappeared before him; or the better negro races may prevail and form nations and exclude the rest."[71] Galton, of course, contradicts himself. In the first part of his definition he defined the negro as a race, but then wrote about "the better negro races." Are we then to assume that Galton meant that "race" can be used to define a whole group as well as its subgroups? Are there better and worse races within a single race? In *Inquiries into Human Faculty* Galton tries to address these contradictions, what he calls "our ethnological ignorance," noting "the absence of a criterion to distinguish between races and sub-races . . . makes it impossible to offer more than a very off-hand estimate of the average variety of races in the different countries of the world." Galton suggests that "on the average at least three different recognised races were to be found in every moderately-sized district on the earth's surface."[72] Galton goes on to define races in a surprisingly heterogeneous way, one that seemingly contradicts the absolute terms in which he spoke about blacks and Africans more generally. For example, on the diversity of races in Africa he writes "that an invasion of Bushmen drove the Negroes to the hills. . . . Then an invasion of a tribe of Bantu race supplanted the Bushmen, and the Bantus, after endless struggles among themselves, were . . . pushed aside . . . by the incoming Namaquas, who themselves are a mixed race. This is merely a sample of Africa, everywhere there are evidences of changing races."[73] But just as surely as Galton wrote of the diversity of races in Africa, he could write just a few pages later that the industrious "yellow races of China" would become a colonizing force in Africa and "extrude hereafter the coarse and lazy Negro from at least the metaliferous regions of tropical Africa."[74] As much as Galton redefined race in a hereditary context, the nature of human diversity precluded him from providing a clear and consistent definition for race itself. Eugenics as a whole suffered from this problem—eugenic literature is rife with multiple definitions and explanations for the nature and meaning of race and how eugenic policy should address these differences.[75]

For Galton, race improvement was "so noble in its aim" that it rose to the level of "religious obligation."[76] By proposing methods for breeding a better race, Galton sought just that. Another of Galton's lasting contributions to the study of race was his proposal to use twins to understand hereditary differences, a research program he began in earnest in 1875.[77] Galton outlined his method for twin studies in *Inquiries into Human Faculty* in 1883, proposing that twins offer "means of distinguishing between the effects of tendencies received at birth, and of those that were imposed by the special circumstances of their after lives."[78] By studying twins who

were "closely alike in boyhood and youth" to "learn whether they subsequently grew unlike, and, if so, what the main causes were," and, con-
versely, to study "the history of twins who were exceedingly unlike in
childhood, and learn how far their characters became assimilated under
the influence of identical nature," Galton hoped to show the primary
impact of nature (versus nurture) on humanity.[79]

# 2

# CHARLES DAVENPORT AND
# THE BIOLOGY OF BLACKNESS

I f Francis Galton was the theoretician of eugenics, then Charles Davenport was its engineer and American torchbearer. In the United States, from the turn of the twentieth century until his death at age seventy-seven in 1944, Davenport was both the public and academic face of eugenics.[1] Through his writings, speeches, and indefatigable advocacy on behalf of eugenic doctrine, Davenport established himself as the doyen of the American eugenics movement. Though Davenport lived just long enough to witness his field in decline in the United States (and its horrific successes on the European continent), during his lifetime he oversaw an expansion of eugenic ideas into both social and scientific spheres that are salient still today.

A New Yorker by birth who prided himself in his colonial birthright, Davenport was the last of eleven children. His father, Amzi Davenport, a real estate man, traced his family line in a multivolume genealogy to eleventh-century England but was more recently descended from Congregationalist ministers in both England and New England.[2] Davenport's love of science was imparted by his mother, Jane, so much so that he dedicated his first book to her, writing an inscription that read, "To the memory of the first and most important of my teachers of natural history—my mother."[3] Davenport's passion for science took him to Harvard, where he earned his Ph.D. in biology in 1892 and afterward became an instructor there. He departed Harvard for the University of Chicago in 1899, taking a position as an assistant professor, where he stayed on until the opening of the Cold Spring Harbor Laboratory on Long Island in 1904. Davenport's success in persuading the Carnegie Institution of Washington to endow

Cold Spring with ten million dollars as a laboratory for the study of evolution marked what one biographer noted was Davenport's early signs
"of being an energetic organizer."[4] This attribute would help Davenport
prosper throughout his long career.

Davenport is most often remembered for seeking to quantify the heritability of social and mental traits as they varied by group, and for the way
in which he linked questions of heredity with the challenges of immigration. As the director of the Cold Spring Harbor Laboratory for almost
forty years, Davenport built the infrastructure of American eugenics at his
beachfront laboratory on the north shore of Long Island. From there he
oversaw an expansion in the study of "the germinal differences that affect
not only form and color and details of physical features but also instincts
and temperament." In a speech in 1920, Davenport warned, "It is not sufficient that a community or a state should purge itself of the 'inferior
strains.' It must guard itself against the immigration into the community
of persons carrying bad germ plasm."[5]

In the years before the passage of the anti-immigrant Johnson-Reed
Act of 1924, which radically limited the entry of legal immigrants into
the United States, eugenicists like Davenport and his colleagues Madison
Grant, Lothrop Stoddard, and Harry Laughlin stoked anti-immigration
fervor by intensely lobbying public and political interests with eugenic
rationales for a closed-door policy. American nativism would reach a peak
in the 1920s, declining once the doors to most immigrants were firmly
shut in the wake of Johnson-Reed. The rise of anti-immigrant fervor in
the United States dated back to at least the 1880s, which saw the beginnings of an explosion of immigration from southern and eastern Europe.
The challenge to the old order, brought about by these new immigrants,
would stoke nativism and anti-immigration fervor for decades to come. In
1882, for example, 1.2 million souls turned up on America's shores, more
than 80 percent of whom had begun their journeys in southern or eastern Europe, prompting one nativist to worry that America was allowing
"every nation to pour its pestilential sewage into our reservoir."[6]

It is no coincidence that the rise of eugenics corresponded closely with
the social transformations brought about by immigration, both real and
perceived, that occurred during these times. By the early years of the
twentieth century, the work of eugenicists would offer scientific rationales

and solutions to the fears of a white elite that was concerned about its and the nation's decline. In the 1920s, for example, Henry Fairfield Osborn, president of the American Museum of Natural History and an ardent eugenicist, lobbied Congress to pass sweeping immigration restrictions. In May of 1924, for example, just a few weeks before the passage of the act, Osborn wrote to Congressman Albert Johnson, cosponsor of Johnson-Reed, congratulating him on the success of the bill in committee, which he believed to be "one of the most important steps taken in the whole history of our country." To Osborn and other supporters, this legislation would spread "throughout the country appreciation of the sacred duty of every American citizen to maintain the character of our country though elevating the character of its people."[7] Osborn had been in contact with Johnson on immigration matters for at least five years prior to the passage of the 1924 act, and his thinking on this matter changed over time. In a December 1919 exchange with nativist and eugenicist Madison Grant, Osborn chastised Grant for his narrow views on immigration, which included quotas by trade and occupation. Osborn suggested instead that "the men, women, and children whom we desire to admit to this country will be tested by their character, their physical health, their willingness to work and their unqualified loyalty." For those who were disloyal to the United States, Osborn proposed immediate deportation. Furthermore, Osborn opposed limiting entrance to select races and proposed instead that eugenics standards could be successfully applied to "men of whatever race."[8] Just a few years later, Osborn's ideas began to take on a nativist tone, urging that a sentinel force be used to prevent undesirables. Osborn wrote to Congressman Johnson in 1922, "My ideas about the future of America are derived not from reading but from personal observation. . . . I think that there are good and desirable immigrants to be found in every country. But all these countries are now striving to keep the desirable people at home, and are sending the undesirables, especially the Jews, to America. This is why it would pay the United States to have observers at all consulates abroad."[9]

The impact of the push to integrate eugenic theory into American immigration policy by Osborn and others was considerable, and the consequences of this pursuit had measurable damaging effects on both immigrants to the United States and eventually on those who died in the

Nazi genocide against the Jews in Europe. Federal immigration restrictions were, as such, buoyed by eugenicist sentiment. Harry Laughlin, the superintendent of the Eugenics Record Office at the Cold Spring Harbor Laboratory, appeared before Congress several times in the early 1920s promoting the belief that immigration was foremost a "biological problem." As Davenport's number two at the laboratory, Laughlin fervently promoted the eugenics cause, maintaining, for example, that recent immigrants from eastern and southern Europe were afflicted "by a high degree of insanity, mental deficiency, and criminality."[10] In his testimony before the House Committee on Immigration and Naturalization, Laughlin pleaded with the congressmen to restrict immigration so America would be allowed to "recruit and to develop our racial qualities."[11] Laughlin believed that the "American Race"—a race of white people from northern and western Europe—was being polluted by a high rate of "inborn social inadequacy" from new immigrants. To be sure, eugenics was not the only catalyst behind immigration restriction legislation. Nationalistic fervor, anti-Semitism, and a more general anti-immigration predilection all combined to advance the restrictive legislation. But eugenics was, in many ways, the most compelling ideology generating support for the bill, particularly because the anti-immigrant prejudices of many in Congress were consistent with eugenic pronouncements on the subject. Science has been a source of tremendous authority in twentieth-century American social and political thought. Scientific claims were used by activists to either bolster or discredit reform efforts. Eugenicists worked hard to capitalize on the rhetorical power of science and its authority in their claims about immigrants.[12] Harry Laughlin, for example, during his congressional testimony in 1923, backed his eugenical vision with what he called scientific facts and criticized attacks on his work as being biased "because its conclusions are displeasing."[13]

Davenport was involved in lobbying on behalf of immigration restriction. In correspondence with Congressman Johnson, cosponsor of the 1924 immigration act, Davenport paid careful attention to the impact of the Johnson-Reed Act on black immigrants from other parts of the Americas. "Could you tell me in a word," Davenport wrote, "whether there is some treaty requirement that makes it necessary to admit Negro immigrants from the West Indies, Brazil and other parts of the Americas. . . .

I believe you will recognize the undesirability of admitting thousands of alien Negroes from any source."[14] Johnson replied, "To date, Negroes have been admitted because of the wording of the Naturalization Act which admits to naturalization free white persons and African Negroes."[15]

The relationship between eugenics, race, and immigration in the 1920s was complex, and scholars have examined how the Johnson-Reed Act, in addition to effectively closing America's immigration doors, also played a role in reconceptualizing racial categories.[16] This redrawing differentiated and ranked Europeans based on their "desirability." Non-European immigrants, including Japanese, Chinese, and Mexicans, were considered unassimilable into American society.[17] Some scholars have suggested that following the passage of Johnson-Reed, the decrease in European immigrants post-1924, along with the migrations of African Americans from the Deep South to the upper south, north, and west, forced black-white racial issues to the forefront.[18]

While eugenicists post-1924 did systematically pay more attention to black-white differences, the idea that the Johnson-Reed Act produced a new racial binary between black and white both overstates the racialized status of white ethnics and fails to acknowledge the longer history and power of the black-white racial binary throughout American history.[19] A new racial alchemy certainly developed during the twentieth century— influenced to varying degrees by social, cultural, political, and economic changes—but this binary was not novel to the twentieth century, nor did it take on a radical new form post-1924. Indeed, evidence from legal history shows that the black-white binary was legally maintained long before the passage of Johnson-Reed, and even long before the consolidation of segregation through passage of Jim Crow laws in the 1890s. An 1851 ruling in Alabama, for example, defined an individual as black "as long as one of his or her great-grandparents was a 'Negro.'"[20] The "Redemption" of the South in the late 1870s and 1880s marked a postemancipation transition to reducing the rights of black citizens. Following the withdrawal of federal troops from the South, the nation abandoned the attempt to guarantee freedmen and -women their civil rights and relinquished control of the political infrastructure to white southerners.[21] The Jim Crow laws of the 1890s legally consolidated the disenfranchisement of African Americans. Furthermore, evidence from the Commonwealth of Virginia shows how eugenics was put in service of the black-white divide as early as 1900, and continued in practice through at least the 1950s.[22]

In the second decade of the twentieth century much of Davenport's research in this area suggested a scientific curiosity about the nature of human diversity, albeit driven by a desire to flesh out these differences in a biological context. But beginning in the latter part of that decade there was a fundamental shift toward asking social questions about human difference, and, as he did in the context of debates about national origin and immigration, Davenport also came to these research questions with the hope of influencing national and regional political debates. Maybe this change would occur in Davenport because the scientific question concerning the genetic nature of skin color would be satisfactorily answered for the eugenicists, and Davenport and others could instead focus on the policy implications of their studies. Or, maybe, as concern diminished in the wake of Johnson-Reed about the eugenic impact of white ethnics on American society, Davenport and others could dedicate more of their resources to the black-white racial binary.

Either way, Davenport's fascination with black-white differences appeared early in his writings, and throughout his long career he struggled with the nature of human variation in a biological context. An undated address titled "A Biologist's View of the Negro Problem," probably delivered sometime in the second or third decade of the twentieth century, confirms Davenport's social prejudices toward blacks in his application of eugenic doctrine to racial difference. "For the present in North Carolina, I am informed these advantages are designed for white persons only," Davenport wrote referring to eugenics, "but for the sake of the progress of society, that socially inadequate persons with darker skin also shall be segregated and kept in happiness but kept from reproducing their kind." "Pity the people that has to depend on laborers, operators, and domestics upon the feebleminded!" he lamented. Finally, he concluded that "as the worst grade is eliminated and a higher grade takes their place at the bottom of the social scale, we shall have our unskilled labor and housekeep service performed by persons who are able to give better service. . . . Thus, then, by eliminating the undesirable and socially inadequate strains without rigard [sic] to color the state may hope to rise an efficiency, morality, and the happiness of the whole people."[23]

But Davenport's views of blacks and race went far beyond this straight-forward application of eugenic doctrine to segregation. From early in his career, Davenport's attitudes toward black Americans were much more than conformity to the racism of the times; indeed, a significant portion of his eugenic research program would examine black-white differences.[24] As early as 1906, Davenport was considering the nature of these differences, particularly in the context of offspring between whites and blacks, and between those whom he repeatedly refers to as mulattoes. In a November 1906 letter written to Aleš Hrdlička, the first curator of physical anthropology at the United States National Museum (which became the Smithsonian), Davenport sought "information concerning the offspring of two mulattoes." He was interested in both the skin color produced by this mating but also wanted to know if mulattoes were fertile and if their children were "vigorous."[25] Hrdlička's response seems to have whetted Davenport's appetite for research in the area. "The question of mixed-blood of whites and Negroes and of their progeny," Hrdlička wrote, "still awaits scientific investigation."[26] Throughout his long career Davenport would repeatedly revisit this line of inquiry.

Davenport's first known published effort to explore the issue of black-white differences came in 1910 in two articles that appeared in the *American Naturalist*. Cowritten with his wife, Gertrude, a former graduate student in zoology, "Heredity of Skin Pigmentation in Man" sought to "provide a more extended, less biased treatment" of "the phenomena of inheritance of human skin color."[27] The essay is written, save one short passage, as a dispassionate and objective exploration of this research problem. It is heavy on data, with numerous figures, and concludes that "skin color in negro X white crosses is not a typical 'blend' as conceived by those who oppose the modern direction of research in heredity, but that . . . the original grades of heavy and slight melanogenesis segregate in the germ cells."[28] The Davenports' conclusion that skin color was a Mendelian trait that segregated was part of an effort by eugenicists and geneticists to lay claim to unanswered questions about the nature of human heredity. Historian of science William Provine suggests that as Mendelians helped uncover mechanisms of heredity through plant and animal crosses, they believed that the laws of inheritance could also be studied through their experimental work, including "an objective appraisal of race mixture in humans."[29] The Davenports' study was the first serious effort in this area by Mendelians.

There are three pieces of evidence, however, that suggest that the Davenports' interest in this research problem was also motivated by their social views, by the racial mores of the times, and, ultimately, by an intent to further the development of the modern science of race and racial difference. First, by "showing" that skin color was a Mendelian phenomenon, the Davenports and other eugenicists and racists biologized race in a modern genetic context. Their paper is one of the earliest examples of a racial characteristic coming under the scrutiny of genetics research and, as such, was a direct intellectual descendant of Galton's research program concerned with redefining race in a hereditary and biological context. If skin color followed Mendel's laws, then, as Davenport would argue in future papers, other traits associated with race must also be genetically heritable. Furthermore, in making a case for Mendelian segregation, Davenport also provided fodder for racial segregation. The Davenports' evidence that blacks who were light skinned would have offspring that "will show the various intermediate grades due to diverse combinations of the black and white units" confirmed racist beliefs (backed up by biological laws) about the indivisibility of blackness. Genetics thus supported the rationale behind segregation and antimiscegenation laws.[30] Eugenicists Paul Popenoe and Rosewell Hill Johnson wrote in support of antimiscegenation laws in their seminal text *Applied Eugenics*, calling these laws "desirable." The authors worried that the "disharmonies" produced by interracial matings would produce offspring that "will usually be inferior to those resulting from a better-assorted mating."[31] In correspondence four years before the publication of *Applied Eugenics*, Davenport wrote to Popenoe highlighting his findings that skin color was a Mendelian trait.[32]

Second, the data analyzed by the Davenports was provided by H. E. Jordan, a noted eugenicist and racist and professor of histology and embryology at the University of Virginia. In his research Jordan addressed "the eugenic aspect of the Negro question" and suggested that the "pure Negro" was inferior to the mulatto because of mulattoes' white blood. He determined to solve "our Negro problem" by "conserving the present mulatto stock and employing it as a leaven in lifting the colored race to a higher level of innate mental and moral capacity."

The introduction of white blood would (somehow) predispose these off-spring to "know their place."[33]

Davenport and Jordan corresponded for nearly ten years, from 1906 through 1914, on a variety of eugenic topics, including their work on what Jordan repeatedly referred to as "the study of the Negro problem."[34] In a July 1913 letter to Davenport, Jordan describes the receipt of a Phelps Stokes University of Virginia fellowship to study this "problem" and seeks Davenport's cooperation in this area of work. In an August 1913 letter, Davenport agrees, writing that it "would be a fine thing if we could dem-onstrate the heredity of such matters as serve to differentiate the two races; such as the alleged difference to resistance to cancer; to tuberculosis . . . to educability; to sex control. Then there might be something definite to say about the consequences of miscegenation."[35]

The Davenports never made their own explicit claims in the *American Naturalist* paper about the relationship between the social and physical traits of those under study. They instead let Professor Jordan express those connections. In the article the Davenports introduced Jordan's observations and data, writing, "We may now consider the pedigrees of skin color collected by Dr. Jordan." Quoting Jordan, they described the "pedigrees" of his study subjects: "One man is a minister, one principal of the colored school, one a thriving merchant, and one a barber, and all seem considerably above the grade of morality and intelligence of the ordinary stupid and irresponsible Negro." In private correspondence a year later, Jordan proposed to Davenport a study of "the relative mental capacity of Negro, mulatto and white school children," asking him, "Don't you think this is a very important field of work?"[36] In his reply Davenport concurred, and he would go on to address this question nearly fifteen years later in *Race Crossing in Jamaica*. The Davenports concluded their discussion of Jordan's data and deductions by telling the reader that "those who know Dr. Jordan will appreciate the better the great weight to be given his conclusion."[37] The Davenports, after all, had to have been aware of Jordan's position on the nature of mulattoes and of his pioneering work bridging the worlds of eugenics, science, and racism. Trading on Jordan's reputation and expertise in matters of race, science, and society bolstered the Davenports' own credentials in the scientific study of black-white differences. The *American Naturalist* paper was one of the first papers published in the twentieth century conflating science and sociology in the study of race.

Third, and finally, Davenport published widely on the nature of black and white mating throughout his long career. Davenport followed his initial foray into the study of race with the publication of his first book on the subject, *Heredity of Skin Color in Negro-White Crosses*, published in 1913 by the Carnegie Institution of Washington. The study, funded by E. H. Harriman, who several years earlier had endowed the Eugenics Record Office at Cold Spring Harbor, was conducted in Jamaica and Bermuda. Davenport's conclusions were nearly identical to his earlier work in this area, though he did pay more careful attention to several of the issues mentioned in the *American Naturalist* essay, including "passing" for white by individuals with "a certain dilution with white blood" and the "matter of great social moment" of "the possibility of a reversion in the offspring of a white-skinned descendant of a negro to the brown skin color."[38]

In a 1917 essay published in the *Proceedings of the American Philosophical Society* titled "The Effects of Race Intermingling," Davenport again addressed the challenges of studying race, and also articulated policies to deal with America's racial diversity. Davenport's definition of race is especially worth noting, for it was, like Galton's, both confusing and inconsistent. Davenport offered what he called a modern geneticists' definition of race—that is, "a more or less pure bred 'group' of individuals that differs from other groups by at least one character, or, strictly, a genetically connected group whose germ plasm is characterized by a difference, in one or more genes, from the other groups." Davenport's example suggests that every small group of individuals who shared similar traits was its own race. He cites "the blue-eyed Scotchman" who belongs to a "different race from some of the dark Scotch." (Despite this distinction, Davenport refers to the race of Scotch just a page later, prompting the reader to consider whether he is now conflating the blue-eyed and the dark Scotch or is still maintaining them as distinct groups.) Davenport even goes as far as suggesting that "as the term is employed by geneticists," racial groups "may be said to belong to different elementary species."[39] This definition, if applied willy-nilly, as its spirit suggests, would make it impossible to quantify difference between any group. The consequences of this approach highlight Davenport's inability to use the race concept consistently and with discrete meaning. Davenport's own studies of the heredity of skin color that recognized that traits were not fixed in a race contradicted such assertions.

"The Effects of Race Intermingling" also highlighted what Davenport believed to be the greatest danger to the sanctity of races—miscegenation. In Davenport's view the offspring of such relations would, "despite the great capacity that the body has for self adjustment," fail to "overcome the bad hereditary combinations." Davenport believed that the hybridization of races was a threat to the social and political fabric of America, so much so that Davenport worried it would weaken the nation. Only through a program that restricted immigration, promoted "selective elimination," and sponsored "eugenical ideals" (i.e., acknowledging that "miscegenation commonly spells disharmony") would "our nation take front rank in culture among the nations of ancient and modern times."[40]

Into the 1920s Davenport continued to struggle with a "scientific" definition of race. In a 1921 lecture he affirmed the role that scientists would play in developing this definition, writing that the "larger popular interest in races and racial traits now more than ever before" necessitated "a reconsideration of the idea of race and the definition of particular races." In that same speech Davenport illustrated the eugenic consequences of what he referred to as "hybridization between a dominant and a subordinate race" and, as such, offered scientific legitimacy to the development of antimiscegenation laws that were, in the 1920s, being passed by southern legislatures.[41] By 1929, when Davenport cowrote *Race Crossing in Jamaica*, his opus on race and genetics, his research had confirmed for him what he had been moving toward for almost twenty years, that there existed "fundamental differences in mental capacity between . . . Negroes and Europeans."[42]

In the Jim Crow era there were manifold ways of enforcing America's racial customs, and the diffusion of eugenic thought into laws across the American South was complemented by the dissemination of eugenic ideas into extralegal racist organizations such as the Ku Klux Klan. It was discovered in 1923, for example, that the historian and eugenicist Lothrop Stoddard was advising the Ku Klux Klan on matters of race. Stoddard's bile-filled popularization of eugenic rhetoric, *The Rising Tide of Color Against White World-Supremacy*, argued that the colored peoples of the world—yellow, brown, and black—were on the verge of overwhelming white supremacy through rapid population growth, the demise of colonialism, and poor white breeding practices. Stoddard reserved special antipathy for the black peoples of the world, writing that at "first glance we see that, in the negro, we are in the presence of a being differing profoundly not merely from the white man but also from those human types which we discovered in our surveys of the brown and yellow worlds. The black man is, indeed, sharply differentiated from the other branches of mankind." Not only were blacks distinct from other nonwhite groups, they also had "no historic pasts. Never having evolved civilizations of their own, they are practically devoid of that accumulated mass of beliefs, thoughts, and experiences which render Asiatics so impenetrable and so hostile to white influences. . . . The negro, on the contrary, has contributed virtually nothing. Left to himself, he remained a savage, and in the past his only quickening has been where brown men have imposed their ideas and altered his blood." Finally, Stoddard's antimiscegenation rhetoric asserted that "crossings with the negro are uniformly fatal."[43]

In January 1923 Stoddard was exposed by the magazine *Hearst's International* to have been offering advice to, and been a member of, the Ku Klux Klan. The article included reprints of letters confirming Stoddard's relationship with the Klan. In addition, the *Hearst's* article reprinted a letter imploring members of the Klan to read Stoddard's book: "Any white man that reads this book will have the fear of God put into him over the race question. Every Klansman should read it, and be able to quote the high spot." In a letter to Henry Osborn soon after the article's publication, Stoddard described his disdain at "the radical-Jew outfit" who "has determined to 'get me' and discredit me if possible." "Through the theft of some papers from the correspondence files of 'The Searchlight,' the organ of the KKK," Stoddard continued, "they have discovered that I have been advising the leaders of that organization on racial and radical matters."[44]

While Davenport never showed direct support for racist organizations like the KKK, his work confirmed and reinforced what Jim Crow customs had claimed for several decades: that the offspring of interracial relationships always resulted in children who were members of "the subordinate or inferior race." Davenport pointed out that, in the United States, "thus negro-white crosses are generally called negroes" and that "this custom has developed with the aim, more or less conscious, of protecting by this classification the superior or dominant race from legal marriage with the inferior stock—from a dilution of their stronger stock by the weaker traits."[45]

A major exhibit and conference at the American Museum of Natural History in New York City would provide the public with an opportunity to learn firsthand the science of eugenics at the Second International Congress of Eugenics, held at the museum in the fall of 1921. The intersection of eugenics and American racism would be on full display at the congress. Eugenicists hoped that the museum event would be a momentous opportunity for the public to learn about their science.

# 3

# EUGENICS IN THE PUBLIC'S EYE

■■ ▌doubt if there has ever been a moment in the world's history when an international conference on race character and betterment has been more important than the present," said Henry Fairfield Osborn, the noted paleontologist and American Museum of Natural History president, in his opening remarks to the Second International Congress of Eugenics.[1] The congress, held at the museum in September of 1921, was a gathering of prominent American and European eugenicists who came together to promote and popularize a eugenical vision of the world. To Osborn and his heredity-minded colleagues, eugenics was a social movement that could have a profound impact upon human populations through the improvement of genetic stock. Said Osborn, "To know the worst as well as the best in heredity; to preserve and to select the best—these are the most essential forces in the future evolution of human society."[2]

Charles Davenport gave one of the keynote speeches at the congress. In it he reinforced the idea that "not only our physical but also our mental and temperamental characteristics have a hereditary basis."[3] Consequently, Davenport hoped that "the study of racial characters will lead men to a broader vision of the human race and the fact that its fate is controllable."[4] The museum event was eugenics in its prime; it was an articulation of racial theories that, during the first third of the twentieth century, impacted far beyond the narrow confines of the academic circles where eugenics was widely celebrated. A strange mix of proceedings characterized the events at the museum, and ideas about eugenics—both its social and its scientific meanings—were on prominent display.

Presentations at the congress were delivered by such rising stars in genetics as Sewell Wright and L. C. Dunn. Although Wright did present at the congress, he never embraced eugenics. His biographer, William Provine, believes that though Wright "had no theoretic objections to eugenics," he "never published on eugenics" and "steered clear of the subject." Part of Wright's resistance "was his belief that human heredity was very complicated and little understood, giving little scientific basis for a eugenics movement at that time."[5] Dunn briefly flirted with eugenics in the 1920s but quickly dropped any association with the movement, displeased with its methodology and unnerved by its overtly prejudiced mission.

Thomas Hunt Morgan, a father of modern genetics, had been an early participant in the eugenics movement as a member of the Committee on Animal Breeding of the American Genetics Association. Morgan, however, was uncomfortable with the eugenic conception of race, and in correspondence with Charles Davenport, he pointed out that human races share more genes in common than not. In 1915, a full six years before the congress, Morgan officially resigned from the committee, citing "reckless statements and the unreliability of a good deal that is said" by eugenicists.[6]

But the vast majority of those who attended the congress were committed eugenicists who would stay with the movement until its dying days two decades later. Charles Davenport, the director of the Cold Spring Harbor Laboratory, and Leonard Darwin, son of Charles Darwin and president of the Eugenics Education Society of Great Britain, were the two most prominent eugenicist luminaries at the congress. The divide between eugenics and genetics had begun to develop in the previous decade. But eugenicists did make some important advances in the study of human heredity. Alexander Graham Bell, remembered best for inventing the telephone, was at the congress as an exhibitor. At the crossroads of eugenics and genetics, Bell's presentation examined "the relation between age of fathers at death, age of mothers at death and longevity of the offspring." Bell's data was an attempt to examine complex intergenerational traits in a hereditary context. However, it was when eugenicists began making claims about the genetic nature of personality traits, intelligence, and complex social behavior that many geneticists left the movement, realizing, as Thomas Hunt Morgan had six years earlier, that such claims were not based on scientific study but on speculation rooted in social beliefs.[7]

The museum itself figured prominently in the congress. Not only were its president, several distinguished members of its Board of Trustees, and members of its curatorial staff involved in the planning and execution of the congress, but also many in the museum community were fervent eugenicists who considered the meeting "the most important scientific meeting ever held in the Museum."[8] Osborn persuaded the Board of Trustees, with little opposition, to host the congress. The museum's 1921 *Annual Report* highlights their institutional position on the congress: "Inasmuch as the World War left the finest racial stocks in many countries so depleted that there is danger of their extinction, and inasmuch as our own race is threatened with submergence by the influx of other races, it was felt by all present and especially by our foreign guests that the American Museum of Natural History had rendered a signal service in providing for the reception and entertainment of the large number of distinguished men and women who attended the Congress."[9]

Opposition to Osborn and the museum's eugenic position didn't arise until two years later, following the publication in 1923 of the third edition of the eugenic tome *The Passing of the Great Race*, by Madison Grant. Grant, a museum trustee and close friend of Osborn's, raised the ire of another museum trustee, the banker Felix Warburg. Warburg was outraged by both Osborn's laudatory introduction of Grant's book and the anti-Semitism infusing the work. Osborn was a notorious anti-Semite and an active booster of Nazi Germany, once writing to a colleague who had recently resigned from the Galton Society for its increasing anti-Semitism that "the only way to learn the truth about Germany is to spend a summer there and freely mingle with these wonderful people who have so much to teach us." Warburg requested that the Board of Trustees investigate Osborn's anti-Semitism and his work with Grant, and described the ideas contained in *The Passing* as "scandalous" and "shameful." A committee of trustees investigated the matter but decided that Osborn's and Grant's ideas were opinions—not scientific statements—and wrote that "there was no need for anyone to feel offended." Osborn, it should be mentioned, later made an "enthusiastic" trip to Nazi Germany. In 1934 Osborn received an honorary degree at the Johann Wolfgang Goethe University.[10]

The museum, one of the world's leading institutions for anthropological thought, was heavily entangled in the racial debates of the times. Early in the century, Columbia University professor Franz Boas, a former museum curator who went on to international renown as the world's preeminent anthropologist, had attempted to move the museum away from typological racial thought. Unable to initiate the institutional change he sought, Boas left for Columbia University in 1905. During the ensuing decades the museum remained split between the racists and a growing stable of more progressive anthropologists, until the mid-1920s, when it began to move slowly toward Boasian anthropology.[11]

Boas's position on race was not popular at the second congress. Earlier in 1921, in an essay titled "The Problem of the American Negro," published in the *Yale Review*, Boas offered a sharp rebuff of the thinking of the eugenicists and white supremacists who gathered at the museum to discuss eugenics. Where eugenicists and racists like Madison Grant believed that "moral, intellectual and spiritual attributes are as persistent in nature as physical characters and are transmitted substantially unchanged from generation to generation," and that races were the units of this intergenerational legacy, Boas countered that "when we talk about the characteristics of the race as a whole, we are dealing with an abstraction which has no existence in nature."[12] Boas believed the idea of race disregarded "the variability of individuals" and in doing so neglected "the differences from the ideal picture in bodily form and make-up among the persons that compose each people."[13] Boas was at once arguing against racial typology and hereditary notions of racial difference and also maintained that insights in neither biology nor psychology could offer "justification for the popular belief in the inferiority of the Negro race." In Boas's estimation, racial prejudice was social in nature and was "founded essentially on the tendency of the human mind to merge the individual in the class to which he belongs, and to ascribe to him all the characteristics of his class."[14] Upon the publication of the *Yale Review* essay, Walter White, head of the National Association for the Advancement of Colored People wrote to Boas expressing his "hearty and sincere thanks for [the] splendid article." White considered the essay "not only as a genuine scientific treatise of great merit but as a distinct contribution to the problem not of the Negro alone but of the welfare of America."[15]

Early in his career Boas had embraced the biological race concept, although he did so with a bit of an egalitarian streak. In an 1896 lecture called "The Races of Man," Boas acknowledged the then commonly held belief that "the brain of the Negro does not grow and develop as long as that of the white man. In this he is decidedly at a disadvantage." But Boas insisted that "we must not interpret the fact as meaning that the Negro cannot attain a culture such as the one which we now possess." Boas also recognized early on that despite what he believed to be the general superiority of whites to blacks, "there will be a vast number belonging to both races who will be equal in all aptitudes" and that therefore "we must take care not to overestimate the amount of this difference."[16] Just eight years later Boas, invited by W. E. B Du Bois to deliver the commencement address at Atlanta University, offered a changed vision of race difference to the graduates: "To those who stoutly maintain a material inferiority of the Negro race and who would dampen your ardor by their claims, you may confidently reply that the burden of proof rests with them, that the past history of your race does not sustain their statement, but rather gives you encouragement. . . . That there may be slightly different hereditary traits seems plausible, but it is entirely arbitrary to assume that those of the Negro, because perhaps slightly different, must be of an inferior type."[17] Boas spent his career developing alternative theories to the dominant racial discourse, "lines of thought" that anthropologist Lee Baker calls "inimical to the consensus about racial inferiority held by people in the mass media, the academy, southern state legislatures, and each branch of the federal government."[18]

As much as Boas and his antiracist contemporaries sought to utilize science as a force for altering the racialized status quo, racial thinking in the early 1920s remained firmly dominated by eugenics and other variants of racist thought. The proceedings at the second congress at the American Museum of Natural History were an example of this continued dominance in American scientific and social thought. The public's access to this thinking was through a temporary eugenics exhibition held in conjunction with the congress. The exhibit, which ran from September 22 through October 22, 1921, drew between 5,000 and 10,000 visitors, according to museum estimates, and took up space on two museum floors.[19] In the Hall of the Age of Man, where the "principle meetings of the Congress were held," the exhibit focused mainly on "early man and his culture." The design of this segment of the exhibit was simple—it integrated well with the museum's overall approach to natural history and seemed to try hard not to offend noneugenically minded visitors to the congress. The permanent collection on display in the hall included anthropoid skulls and heads. For the congress these specimens were shown in "new positions for the purpose of coordinating and emphasizing geological history of the human species." The temporary exhibit included displays showing "Man's Place Among the Primates" and "The Most Ancient Human Races."[20]

The first-floor exhibition space, held in both the Darwin and Forestry halls, was quite different. There the tools and data of eugenics were on display. The exhibit, as described by attendee Harry Laughlin, "comprised mainly embryological and racial casts and models, photographs, pedigree charts and tables, biological family histories and collective biographies . . . maps and analytical tables demonstrating racial vicissitudes, anthropometric instruments, apparatus for mental measurements, and books and scientific reprints on eugenical and genetical subjects."[21] Eighteen thematically organized booths were filled with 131 exhibits. The attendees were "college and university professors, investigators in scientific institutions, physicians and field workers in institutions for the socially inadequate, statisticians and research departments of the great life insurance companies, scholars and authors of independent means. . . ."[22] Indeed, they were a distinguished group.

Examples of the booths' topics were "Eugenical Organization," "Human Heredity," "Anthropometry," and "Mental Testing, Psychiatry." One booth, designated "Genetics and Heredity," sought to secure genetics at the center of eugenic theory, speaking of it as "an important foundation factor in eugenics or the improvement of the human race through a knowledge of heredity and its application to selection and fecundity."[23]

Race was a theme in all the booths. Two separate booths on the "Races of Man"—booths dealing specifically with race as a black-white issue—were prominent in the conference exhibition. One "race" booth explored "the history of the origin and development of races and the analysis of the . . . determination of the hereditary nature of specific traits." The other "race" booth focused on the "elementary qualities" of races and their role in human progress.[24] Other booths touched on black-white racial dynamics. Dr. Thomas Garth of the University of Texas, Austin, presented his research under the title "Curves Showing Racial Differences in Mental Fatigue." Garth's research showed that black students fatigued more quickly when working on mental tasks as compared with whites.[25]

Another University of Texas at Austin faculty, the zoologist Theophilus S. Painter, presented an exhibit called "The Chromosomes of Man." Painter, one of the world's leading geneticists at the time and a pioneer in *Drosophila* studies, exhibited pictures of chromosomes of a white and of a black man alongside each other so as to show that they were "alike in general form and in number." This photographic display, however, intended something different; it was presented to illustrate a subtle, if not significant, difference between black and white genes.[26] These genetic differences were further highlighted in the presentation by Dr. A. H. Schultz of the Department of Embryology, Carnegie Institution of Washington. Dr. Schultz's exhibit, "Comparison of White and Negro Fetuses," examined the "racial differences during prenatal development of man."[27] Measuring the alleged differences in fetal development between the races, according to Schultz, could visually show just how different blacks were from whites. In the exhibit, finger, leg, thigh, and arm shape and length were all different between the two fetuses. The brain was smaller and the face larger in the black fetuses.

One of the congress's headliners, American Museum trustee Madison Grant, was the vanguard of early twentieth-century racial science. His books sold widely, and his influence was felt beyond the confines of academic eugenics. Author of *The Passing of the Great Race*—one of the most infamous eugenic texts of the era—was a successful corporate lawyer who dabbled in zoology and anthropology and held positions in these fields, including time spent as the head of the Wildlife Conservation Society of the Bronx Zoo. Grant's work as a naturalist propelled him to the forefront of the American conservation movement. He was one of the founders of the Save the Redwoods League in California and successfully lobbied the government to make Mount McKinley into a national park.[28] Grant was also head of the overtly anti-Semitic and anti-Catholic Immigration Restriction League (later renamed the Eugenics Immigration League) and a member of the Galton Society and the Committee on Eugenics.[29] Grant believed his work in conservation and eugenics to be closely related; both were, he wrote in a letter to Osborn, "attempts to save as much as possible of the old America."[30]

At the second congress, *The Passing of the Great Race* was on display as Exhibit 51—"the principal feature of this exhibit consisted of enlarged copies of the several maps which appeared in the exhibitors book."[31] Grant's work, now known for the influence it would have on Nazi ideology, "enjoyed considerable vogue in the nineteen-twenties."[32] It is not hard to see why the Nazis embraced Grant's work. *The Passing* argued that the successes and failures of the civilizations of Europe could be correlated to the amount of a nation's Nordic "blood." Grant surmised that both Europe and America were facing imminent demise because their Nordic bloodlines were fast becoming contaminated by the genetic stock of inferior races. In *The Passing* Grant noted, surely to Hitler's later delight, that "only in a few cases, notably in Sweden and Germany, does any large section of the population possess anything analogous to true race consciousness."[33]

Grant extended his opprobrium to the "darker races" of man. It is in these passages of *The Passing* where the unmistakable rankness of racial theory grounded in emerging eugenic ideas about the biology of race difference can be found. To Grant and his eugenically minded colleagues, it was non-Nordic racial groups in Europe who presented the most immediate threat. Nevertheless, the fear of diluting the white race with black blood was unmistakable. Grant feared black-white racial amalgamation was creating a "population of race bastards in which the lower type ultimately preponderates."[34] But the perceived threat from blacks was not as real as the possibility of the amalgamation of blood from white immigrants. After all, black-white miscegenation was already illegal in most states, and culturally proscribed in the rest. Moreover, Grant deemed blacks incapable of threatening the Nordic race, writing, "Negroes have demonstrated throughout recorded time that they are a stationary species, and that they do not possess the potentiality of progress or initiative from within. Progress from self-impulse must not be confounded with mimicry or with progress imposed from without by social pressure, or by the slavers' lash."[35] Finally, Grant believed that blacks were a "valuable element" in European and American societies, worth being preserved as a "servient race." Any attempt at social equality, he concluded, "will be destructive to themselves and to the whites."[36]

To these pernicious thoughts about blacks Grant applied his eugenic conviction. That is, that "the great lesson of the science of race is the immutability of somatological or bodily characters, with which is closely associated the immutability of physical predispositions and impulses."[37] Grant's race-centered worldview asserted that "race lies to-day at the base of all the phenomena of modern society."[38] This outlook, one that combined a view that accepted race as based in biology, transmitted at the level of the gene, with a belief that racial problems dominate the political and social landscape, was the intellectual force behind eugenics. Its impact on American racial thought was explicit, and its consequences enduring. And its application to black Americans was neither accidental nor indirect. But because he was not a scientist and because his ideas may have been publicly distasteful to some, Grant is often cast as an outsider compared with more "mainstream" eugenicists like Charles Davenport or Harry Laughlin. To the contrary, Davenport and Grant were close allies in the movement, both socially and politically. Davenport worked with Grant on anti-immigrant campaigns and Grant was a funder of and a fundraiser for Davenport's work at Cold Spring Harbor, introducing him to the New York City social fundraising circuit, where, historian Elazar Barkan notes, "bigotry and racism was a popular recreation."[39]

In addition to the exhibition and booth presentations at the congress, 108 papers were presented during the six-day meeting. Presentations were given on a wide variety of subjects, ranging from "The Effects of Inbreeding on Guinea Pigs," presented by Sewell Wright, a leading geneticist, to a paper titled "Individual and Racial Inheritance of Musical Traits" by Carl E. Seashore of the National Research Council. Many of the event's papers confirm that eugenics was not just about race in the context of immigration. These presentations, like many of the booth exhibits, highlighted the emerging new language of a genetic racial science, specifically as it would apply to black Americans. And while many of the papers dealt generally with the race concept, almost 10 percent dealt in a direct way with black-white differences.

R. Bennett Bean, a leading anatomist and physical anthropologist at the University of Virginia, who made seeking to quantify black-white differences in physiognomy his life's work, presented a talk titled "Notes on the Body Form of Man." Bean had published widely on the nature of the relationship between the physical and mental inferiority of black Americans for several decades. Trained at Johns Hopkins under the leading anatomist Franklin Mall, Bean quickly established himself as a leading authority of black and white anatomical differences and published widely on the subject. Through a series of papers during the first decade of the twentieth century, Bean argued that black brain structure revealed significant differences between blacks and whites in cognitive ability.[40] Bean believed that these differences made blacks and whites "fundamentally opposite extremes in evolution."[41] *American Medicine*, a leading medical journal, editorialized in support of Bean's thesis, writing that "no amount of training will cause that [black] brain to grow into the Anglo-Saxon form," and suggested that his studies proved "the anatomical basis for the complete failure of Negro schools to impart the higher studies." In 1909, Franklin Mall, Bean's medical school mentor, sought to verify Bean's measurements of black and white brains, and could find no significant differences between black and white brain structures. "I have now had considerable experience in the dissection of the Negro and have yet to observe that variations are more common in the Negro than in the white," Mall wrote in a rebuttal in the *American Journal of Anatomy*.[42] Still, Bean's ideas about racialized anatomy quickly became the scientific and popular norm, while Mall's work had little impact.[43]

In "Notes on the Body Form of Man," presented at the second congress, Bean compares stature and sitting height in groups he labeled American whites, Negroes, and Filipinos in an attempt to show a hierarchy of physical differences between races. According to Bean's data and analysis, blacks are consistently "discernible by reason of the lower index." Bean includes a long description of the striking physical differences between blacks and other groups and describes what he believes is "the true Negro," a person having "a shortened torso and relatively long legs." The data presented, however, do not represent this "true Negro," since "the records are of Africans with a large admixture of European blood."[44] It is on this point, at the intersection of physiology and blood, that Bean's typology becomes eugenic in nature and is thus used to suggest the permanence of racial bloodlines.

Several other presentations at the congress offered fresh support for a geneticization of the black-white divide. The distinguished anthropologist E. A. Hooton presented a paper titled "Observations and Queries as to the Effect of Race Mixture on Certain Physical Characteristics."

Hooton, who spent his entire teaching career at Harvard University and was associated, but never closely affiliated with, the eugenics movement, wrote about the morphological differences between major racial groups and "observed" the effects of racial hybridization in crosses between blacks and whites.[45] "Distribution and Increase of Negroes in the United States," by W. F. Willcox, an economics professor at Cornell, and "The Problem of Negro-White Intermixture and Intermarriage," by Frederick Hoffman, an actuary for the Prudential Insurance Company of America and author of the racist tract *Race Traits and Tendencies of the American Negro*, explored the demographic trends of African Americans. Willcox claimed in his essay that while in the South the birth rate for blacks exceeded their death rate, in the North the reverse was true. Thus, the increase in the northern black population was due entirely to immigration.[46] Willcox believed that urbanization was killing America's blacks at a high rate, and that their rate of growth as a population would decline rapidly, to the point that "20,000,000 should be accepted as the maximum limit of the Negro population of the United States at the end of the century."[47] Willcox's data and conclusions suggested, as did Hoffman's earlier conclusions in *Race Traits and Tendencies*, that blacks as a group were deteriorating, or as Hoffman put it, "The colored race was on a downward grade," leading toward "extinction," a conclusion the distinguished sociologist and leader W. E. B. Du Bois called "absurd."[48] This decline would be attributed, in both Willcox's and Hoffman's work, to biological, and not environmental factors.[49]

In his speech opening the eugenics congress, Henry Osborn articulated the new scientific language of race and racism. Osborn expressed this emerging ideological consensus on race, saying, "The reason that these races are so stable and maintain their original character so stoutly is that the most stable form of matter which has thus far been discovered is the germ plasm on which heredity depends."[50] Eugenic research throughout the 1920s continued to integrate this idea into its political advocacy, increasingly in the area of black-white difference. The language of science and the language of heredity were integrated into the American zeitgeist to become the intellectual justification behind the pernicious ideology of American racism. In the remainder of the 1920s, with eugenics at its most popular and powerful, the followers of the movement continued the work begun by Francis Galton in the 1860s. Charles Davenport, of course, led the way, helping to lay a foundation for a century of research into black-white differences by eugenicists and geneticists, as well as by other fields drawing on their scientific findings. While Galton laid out the theoretical basis for the relationship between heredity and black-white differences, Davenport developed methodologies to test and measure these differences.

Davenport's connection between his study of black-white differences and his involvement in this issue as a policy maker became only more pronounced during the 1920s. His research into black-white differences in Jamaica, for example, grew into an international research project that, in 1928, was published as *Race Crossing in Jamaica*. The work was funded by Wickliffe Draper, whose interest in racial science would, in the 1930s, grow into his founding of the Pioneer Fund, an organization that still funds racial research today. Davenport and other eugenicists would also continue to research other areas of black-white differences, including a study proposed by the National Research Council to look at the educability of black children as compared with white children by setting up "identical" but separate orphanages to test for racial difference.[51]

The 1920s also saw the beginnings of a comprehensive response to racial science by geneticists, biologists, and anthropologists, as well as by social scientists, who would come to play an increasingly important role in developing research in support of or, alternatively, in challenging the accuracy and assumptions of racial science. At the National Research Council of the National Academies, a series of committees were formed to study race that today provide insight into the evolving race concept and the scientists deeply involved in negotiating its meanings. It is also in these committees that we can see the deepening of eugenicists' interest in the nature of the social and biological differences between blacks and whites, an interest that had been overshadowed by eugenics' attention to immigration and ethnicity.

# 4

# THE NATIONAL RESEARCH COUNCIL AND THE SCIENTIFIC STUDY OF RACE

The final report of the Joint Commission on Racial Problems capped more than a decade of efforts by the National Research Council of the National Academy of Sciences to grapple with the scientific and sociological meanings of race in America. In his final report in 1931, committee chairman and Columbia University psychologist Robert S. Woodworth described a plan to build racial orphanages to study the biological, psychological, and sociological meanings of race in controlled environments. According to the report, this "comprehensive study of child development in different races would be an important contribution to the study of race differences." The orphanages, one white and one black, would provide an "excellent environment, with opportunities for scientific study, organized to receive a sample of negro children, for example, soon after birth, and to retain them for several years at least."[1] Though this study was never funded, it is not coincidental that at the same time the notorious "Tuskegee Study of Untreated Syphilis in the Negro Male" was in its earliest stages (interestingly, a small study in England in the 1970s found that black and white children raised together in an orphanage did, in fact, have different outcomes in IQ scores—the black children's mean IQ was slightly higher[2]). Both studies assumed significant roles for federally sponsored science and both assumed that traits unique to blacks and whites were worthy of study and expense.

During the 1920s the National Research Council convened, in Washington, D.C., a series of investigatory committees to study the biological and social aspects of race difference in the United States and sponsored research to that effect. Beginning with the Committee on Race Characters in 1921 and ending with the Committee on Racial Problems in 1928, four separate National Research Council (NRC) committees brought together prominent geneticists, biologists, eugenicists, anthropologists, psychologists, medical doctors, and others to explore the impact of human difference on the United States' social, political, and economic systems. While the racial groups studied by the NRC changed over the course of the decade (a shift from European ethnics to African Americans), an interest in the way human difference was shaping 1920s America remained strong throughout.

At the outset of the 1920s, the Committee on Race Characters studied the assimilation of European immigrants into American society. In 1921, Clark Wissler, an anthropologist at the American Museum of Natural History in New York and a member of the Committee on Race Characters, wrote in urgent support of funding NRC projects in this area "since the most complex situation confronting our nation today is the assimilation and Americanization of the large and diverse racial groups now present in our population."[3] The Committee on Racial Problems focused instead on race as primarily an issue of black-white differences.[4] This evolution, from the study of white ethnics to blacks, from concerns about immigration's impact on the racial character of the United States to concerns over a demographically changing African American population, is unmistakable when examining the NRC committees' records and indicates both the shifting priorities of a changing nation and continued attention to long-standing antiblack prejudices.

The biological and sociological assumptions about racial difference that drove the work of the NRC race committees during the 1920s were not the only efforts to pathologize black Americans, nor were eugenic and other scientific manifestations of racial science the only components of white supremacist thought. Sociologists and other social scientists also developed nonbiological rationales for the inferiority of African Americans and other racial groups. For example, social scientists claimed the black personality to be damaged, a theory embraced by both sides of the political spectrum. On the one hand, as historian Daryl Michael Scott argues, conservatives utilized findings of a damaged black psychology to "justify exclusionary policies and to explain the dire conditions under which many black people live." On the other hand, liberals "seeking to manipulate white pity . . . used damage imagery primarily to justify policies of inclusion and rehabilitation."[5]

Racist cultural depictions of African Americans also fueled the belief that they were not fit for citizenship. In the early decades of the twentieth century, literary and cinematic nostalgia for the plantation South helped popularize derogatory images of African Americans. For example, Thomas Dixon's novels, which included the best-selling *The Clansman*, surely reflected and influenced many whites' views of blacks. *The Clansman*, later made into D. W. Griffith's 1915 film *Birth of a Nation*, according to historian Philip Dray, "combined technical and artistic brilliance with a controversial rendering of Reconstruction that rehashed many of the most enduring and painful Southern myths about black Americans." President Woodrow Wilson, who had federalized Jim Crow segregation during his administration, hosted both Dixon and Griffith as guests at the White House to screen the film.[6]

African American intellectuals were prominent among those who responded to the growing chorus of racist thinking and action at this time. This same period saw the formation of civil rights groups, including the National Association for the Advancement of Colored People (NAACP) (1909) and the National Urban League (1910).[7] African American intellectuals played a prominent role in responding to antiblack attacks. Kelly Miller, dean of Howard University in Washington, D.C., in response to Thomas Dixon's popular antiblack tirade *The Leopard's Spots: A Romance of the White Man's Burden, 1865–1900*, asserted the common humanity of all peoples, writing that "since civilization is not an attribute of the color of skin, or curl of hair, or curve of lip, there is no necessity for changing such physical peculiarities."[8]

The attention to black-white differences by eugenicists and the consequent changes to ideas about race and black-white differences occurred at a time of considerable upheaval in African American history. A eugenic position on racial matters must be seen in this context. When the historian Rayford Logan wrote in 1957 that the time from the end of the nineteenth century through World War I was "the nadir" of race relations in the United States, he was referring to the disenfranchisement of African Americans across the American South, the rise of the Jim Crow system of segregation, an escalation in racial violence against African Americans, including lynching and race riots, and the migrations of more than one million blacks out of the South and into segregated northern cities.[9] The late nineteenth and early twentieth centuries were difficult and changing times for African Americans and for America's racial mores more generally. Writing in 1920, at the end of this period, a frustrated W. E. B. Du Bois lamented that "instead of standing as a great example of the success of democracy and human brotherhood America has taken her place as an awful example of its pitfalls and failures, so far as the black and brown and yellow peoples are concerned."[10] The advent of Jim Crow laws across the American South in the 1890s institutionalized white supremacy, effectively nullifying whatever progress had been made toward racial reconciliation during the era of post–Civil War Reconstruction.[11]

Huge demographic shifts also affected African American communities and their relationship with fellow citizens. In all, from the turn of the century through the beginning of World War II, more than one million African Americans moved out of the Deep South.[12] Historian Nell Painter writes that the movement of blacks from South to North during the early twentieth century represented "both immigration and freedom" to African Americans. For those who wanted a life other than Jim Crow's, change brought with it hope, but not always opportunity. And this Great Migration from South to North, from rural to urban, from the Deep South to the upper South, and from farmland to industrial, would come to have an important impact on the nation's racial calculus as formerly disenfranchised southern blacks gained the vote in their new northern homes.[13] These migrations surely forced the federal government to take new notice of the black community and helped lay a foundation for a civil rights movement for African Americans.[14]

The National Academies took careful note of this demographic shift and other phenomena it believed fell under the rubric of racial research, and it assigned a committee to investigate their impacts on "interbreeding" between whites and blacks, the fertility rates of black-white racial crosses, the birth rates and death rates of these crosses, and the variables that might be influencing the results.[15]

## THE COMMITTEES' WORK

The NRC was an ideal institution for the bold and large-scale studies that would be proposed, and in some cases funded, through its various committees on race. As a division of the National Academy of Sciences (NAS), the NRC shared in its scientific prestige and benefited from its political muscle. The NRC had its origins in 1916 when the NAS helped coordinate nongovernmental scientific and technical resources with military agencies as the United States prepared for entry into World War I. As early as 1863, the Civil War had "brought about the organization of the National Academy for the assistance of the Government."[16] A May 11, 1919, executive order by President Woodrow Wilson officially incorporated the National Research Council as a division of the NAS. The articles of organization of the NRC state its purpose as follows: "To promote research in the mathematical, physical and biological sciences, and in the application of these sciences to engineering, agriculture, medicine and other useful arts, with the object of increasing knowledge, of strengthening the national defense, and of contributing in other ways to the public welfare."[17] In addition to its study sections and committees, postdoctoral fellowships and grant funding helped round out the work of the NRC. Between 1919 and 1933 the NRC funded approximately 850 fellows.[18]

The work of the NRC was underwritten both by government funds and by significant monies from the Carnegie and Rockefeller foundations, institutions that also funded eugenics and other forms of racial science at the time.[19] As historian Roger Geiger has noted, these relationships underscored the NRC's role as the nexus "for private, elite authority over American science in the 1920s."[20] To this group—a gathering of "the elite of American science, the heads of the philanthropic world, research directors and corporate leaders of the major research-based firms of the day, and certain key figures from public life"—the direction of American science under the administration of the NRC was entrusted.[21]

The importance of philanthropy to eugenics was recognized in the earliest days of the movement. Just after the turn of the century the Harriman estate was the first to fund Charles Davenport's exploits at Cold Spring Harbor. At a 1914 conference sponsored by the Kellogg family, the relationship between philanthropy and eugenics was highlighted. At the conference, Leon J. Cole, a geneticist at the University of Wisconsin, presented a paper with the title "Biological Eugenics: Relation of Philanthropy and Medicine to Race Betterment" that called for special study of "what medicine and philanthropy are doing for the race" and endorsed studying this research problem "from our knowledge of general biological laws."[22] Cole, in whose name an endowed chair lives on at the University of Wisconsin, recognized the importance of philanthropy in furthering the integration of eugenics into the American landscape. And while he understood that "it might seem odd that philanthropy and medicine should be classed together," he believed that they had the following in common: "The one tends to relieve the want, the other the suffering, and both often to prolong the life of the recipient."[23]

The Carnegie Institution's interest in issues of race and eugenics dated back to the turn of the century. Davenport's first major eugenic project, the Station for Experimental Evolution at Cold Spring Harbor, also known as the Department of Experimental Biology of the Carnegie Institution of Washington, received the foundation's financial largesse and institutional support beginning in 1903.[24] In 1913, the Carnegie Institution published Davenport's first book on black-white differences, *Heredity of Skin Color in Negro-White Crosses*.[25] Finally, the Carnegie Institution funded the Eugenics Record Office, for which Carnegie provided $474,014 for its operating budget between 1918 and 1939.[26]

At the turn of the century, knowledge was quickly becoming, like land and capital, an important resource in the American economy.[27] The Rockefeller and Carnegie foundations both backed the development of racial science through what historian Ellen Lagemann has called "the politics of knowledge."[28] Through their ability to "exercise significant influence on decisions about public problems," determining how and what information "experts would communicate with nonexperts," and by controlling "who could and would gain entrance to the knowledge-producing elites that emerged and proliferated as the United States became a more nationally and bureaucratically organized society," in Lagemann's analysis, wealthy foundations became a vehicle to "define, develop, and distribute knowledge."[29]

While the foundations benefited many, there were also those who expressed concern about concentrating control of knowledge in the hands of a few powerful groups. Much in the same way that monopolies in the production and distribution of oil and steel were attacked, critics also assailed the construction of "cultural empires" in the world of ideas.[30] Still others have suggested that philanthropy has been the "American equivalent of socialism," intending to relieve class tension.[31]

In the case of eugenics and the funding of racial science, however, the impetus was less about relieving class and racial tensions than about solidifying them.

It should come as no surprise then that having played such a significant part in the development of eugenics, these foundations, especially Carnegie, both believed in and sought to preserve white supremacy.[32] The funding of eugenics and other racial scientific initiatives through the NRC was a significant way for foundations to support and bolster this point of view. Beginning in 1917 the Carnegie Corporation provided significant grants toward NRC activities.[33] In August of 1919, the Carnegie Corporation approved a $5 million grant to be a permanent guarantee of income "for general expenses [that] would make the Council a going concern, and enable it to secure additional contributions."[34] That same year the Rockefeller Foundation promised $500,000 for five years for postdoctoral awards funded from the NRC.[35]

The Division of Anthropology and Psychology, formed in October 1919, undertook all racial research at the NRC during the 1920s. In a description of the NRC's mission, the Division of A&P (as it is referred to in NRC correspondence), was directed to coordinate the research activities at the NRC in related fields, to train students in these fields, to foster research in anthropology and psychology, and to "act in an advisory capacity on research projects within our fields, when such counsel is requested by duly constituted agencies." That racial research took place under the direction of this division rather than the Division of Biology and Agriculture seems meaningful only in the context of the research generated by the two committees: the Division of Biology and Agriculture, organized officially in 1919, grew out of an NRC committee on agriculture, and thus most of its research was focused on agricultural biology. Research in marine biology, as well as a project on radiation biology, also fell under the aegis of the Division of Biology and Agriculture, whereas human biological and sociological research fell under the control of the Committee on A&P.[36]

It would not be fair to conclude that the racial research projects at the NRC during the 1920s themselves altered the ways in which scientists and the public understood the biology of human difference. Indeed, the committees' work more likely reflected a growing consensus among mainstream scientists that genetics and biology could be used to test prevailing social ideas about difference. However, that this line of work took place under the imprimatur of the National Academies certainly amplified its impact for three reasons: because of the National Academies' status in American science and politics, because the NRC committees funded racial scientific research, and because the NRC became a hothouse for emerging ideas in the area of racial science throughout the decade.

Two A&P committees focused on the study of race in America during the first half of the 1920s: the Committee on Race Characters and the Committee on Human Migration. In 1923 A. E. Jenks, the chairman of the Committee on Race Characters and a professor of anthropology at the University of Minnesota, wrote to Robert M. Yerkes on Yerkes's acceptance of membership on the committee. Yerkes, a comparative psychologist, is best remembered for his primatology research and for the development of U.S. Army mental tests used "for the classification of men in order that they may be properly placed in the military service."[37] Although the army limited its use of the tests after World War I, this testing was a watershed, opening the door to mass IQ testing in schools and business and shifting the public's perception to one that embraced such testing as an accurate measurement of innate intelligence.[38] The army tests were used in defense of racial segregation and were also cited in support of immigration restriction in the early 1920s.[39]

In his letter to Yerkes, Jenks identified the "need for scientific knowledge (about races) for the benefit of the national strength and well-being," and believed that his committee was poised to "make contributions" toward that end. Jenks offered his a priori assessment of racial difference to Yerkes, admitting that he believed "the so-called races greatly differ" in ways that were primarily genetic in origin. Jenks also did not identify specific racial groups in his assessment of group difference. Rather, he was interested, as the committee's funding patterns would later show, in a general understanding of the racial matters of the nation. Jenks outlined six "measurable matters" by which races differed: (1) "physically as breeding animals, due to gametic differences"; (2) "physically in susceptibility to certain diseases"; (3) "temperamentally, due, perhaps, to differences of balance between the action of the glands of internal secretions"; (4) "psychically in pathologic reactions, due originally, perhaps, to unhappy racial experiences"; (5) "both physically and psychically in capacities and

attitudes . . . due to heredity, and to geographic, human and cultural environment, etc."; and (6) "in America, psychically in instinctive and often
also in deliberate racially-normal reactions to many fundamental phases
of characteristic American life, due to such varied things as history, content of education, ideals, etc."[40]

While Jenks's description of the nature of race difference offered a variety of explanations, ultimately it was biological and genetic reasons that
would receive the most attention by the committee. In the foreword to
the committee's "Research Outline" of 1923, the committee laid out its
goal: "It should be the purpose of racial researchers to arrive at the facts
as to the existing race traits, to measure the traits of each race studied
so that in due time it may be known what characteristic strengths and
weaknesses for America the various races possess."[41] Furthermore, the
committee acknowledged that until this point, "though races are exhibited before all eyes, they are not defined scientifically, but almost entirely
subjectively. . . . The result is much confusion, disagreement and often
bitterness."[42]

The committee outlined six research projects, based primarily on the
recommendations of Paul Popenoe, to begin the scientific study of race
difference.[43] The proposed studies, "Study of Normal Race Traits in a
Selected Few Races," "Typical Pure-Blood Races for Research," "Other
Pure-Blood Groups," "Mixed-Blood Groups," "The Old-Line American
Groups," "The Negro," and "The Assembling of Existing Race Data," complemented each other in an attempt to examine a range of races from
an "unbiased observation of facts, from the presentation of the facts in
unevasive language without exaggeration or argument, and from the inevitable conclusions to be drawn from the facts."[44] The proposed studies
were organized to examine race traits as distinctly physical, intellectual,
moral, and social entities. Because the studies were based on the idea that
races shared bloodlines, the methodology suggests that the traits under
study were hereditary (through genes) in origin. The committee's report
implies as much, stating that "it is the intention in these studies of the
normal race traits to produce model studies in race traits, and also in
human heredity."[45]

The studies also reveal a great deal about how race was defined in the
early 1920s. Pure-blood races included "the Mexican race," "the south Italian

race," the Russian Jewish race," and "the Finnish race," while unidentified mixed-blood groups were considered a by-product of racial "hybridization" that in America "is going on in larger and more complex way[s] than elsewhere in the world."[46] The project on "The Negro" acknowledged only a cursory understanding of individuals grouped under that racial banner. Its proposal, for example, stated that "no race group is less authentically and commonly understood in America than the Negro."[47] Project organizers also expressed interest in studying what it called the three classes of blacks: "those of pure African descent," those who shared "Negro and general white ancestry" ("those commonly known as mulattoes"), and "those of accurately ascertained bi-racial parentage—as Negro X Chinese, Negro X Indian, Negro X Swede."[48] The distinction made between mulattoes and biracial blacks would suggest that in the early 1920s race and nationality, at least for the time being, were one and the same.

Nowhere is eugenics mentioned in these proposals. However, the research clearly followed a eugenic paradigm for racial studies, suggesting that eugenic ideas were, in fact, barely indistinguishable from mainstream studies of race. That so many of the scientists involved in the NRC race committees were themselves eugenicists or displayed eugenic leanings suggests that eugenics was an adjunct to most fields in the natural and social sciences. The work of the NRC race committees, their research proposals, funding patterns, and committee composition support eugenics as much more than just a complement to modern anthropological and biological thought. Indeed, eugenic views of race defined how difference was studied at this time. Evidence from the NRC committees illustrates how the role of eugenics was much more fundamental to the 1920s era of racial studies—including anthropology—than has been previously acknowledged. Other books have limited or overlooked the role of eugenics in the construction of racial thought in the United States. This is due, it seems, to the way scholars have previously defined the eugenical impact on racial studies; that is, as a political program that sought to guide human evolution through reproductive and immigration restrictions rather than as a scientific worldview that helped to define the way in which human difference was viewed, studied, and politicized.[49]

Despite the Committee on Race Characters' far-reaching set of research goals, as of 1923 its projects had received no funding. To address this shortcoming, the NRC established the Committee on the Scientific Problems of Human Migration, with Yerkes as its chair. A small grant of $5,000 from the Russell Sage Foundation helped the committee prepare its work, which was loosely defined as the "formulation of a research program on problems of immigration."[50] Subsequent grants, totaling $145,910, funded the work of the committee through 1928, when it was discharged. The Laura Spelman Rockefeller Memorial funded all but $10,000 of the committee's work.[51] Like the Committee on Race Characters, this new NRC committee advocated the importance of a scientific understanding of racial difference, but did so explicitly within the context of race intermixture or racial amalgamation.

The committee claimed that through vigorous scientific research it could "check the premature, shortsighted, impressionistic, and dangerous speculations of certain publicists" on the matter of the biology of race.[52] The targets of such comments were popular writers like Madison Grant and Lothrop Stoddard, whose best-selling books gained notoriety during this period for their racist assertions concerning the nature of racial differences.[53] It is interesting that many of the members of the committee were avowed eugenicists, including most prominently Charles Davenport. That Davenport and other eugenicists issued such a strong statement against racist popularizers seems hypocritical, especially given the intellectual and institutional cover eugenicists like Davenport and Henry Osborn gave to men like Stoddard and Grant throughout their careers. Nonetheless, it was in the interest of academics like Davenport and Yerkes to distance themselves from such extremists when it was expedient, especially if it was done to attract foundation funds and federal dollars to what they wanted to be perceived as legitimate scientific research.

The approach of this committee to the study of the relationship between immigration and race is recorded in fascinating detail in the transcripts of a one-day conference held on the subject of "racial intermixture" February 17, 1923. Among the participants were the psychologist Raymond Dodge, eugenicist and biologist Charles Davenport, geneticist Raymond Pearl, psychologist Robert Yerkes, embryologist Frank Lillie, biologist Samuel J. Holmes, and anthropologist Clark Wissler. The composition of the committee, leaning heavily toward the biological and genetic, again illustrates how questions of race under the NRC A&P banner were framed, in large part, in a biological context. Also of importance from this conference was that despite the committee's intended focus on general race questions, most of its discussion concerned black-white differences.

The three most significant pieces of the committee's discussion are the seeds of what would become Charles Davenport's *Race Crossing in Jamaica*; a detailed outline of Raymond Pearl's studies of race in Baltimore; and a proposal to "carefully prepare and present to the Secretary of Commerce a plan for the alteration and improvement of the tabulation and publication of the fundamental records of births and deaths of this country, to the end that the splendid mass of material which exists and is constantly accruing may be utilized for biological study."[54] The group discussed how "mental, temperamental, and behavior differences in races is particularly important in estimating the effects of race intermixture." The committee acknowledged that "the fundamental problems of race intermixture are not different in principle from those of any other aspect of human inheritance," and that "the entire subject of genetics must be made to contribute its results."[55] Interest in black-white race mixture was justified because it was "the most conspicuous instance of race intermixture in America." The committee recognized that despite "practically no Negro immigration at the present time," "secondary migration" of African Americans "within the nation, as from country to town, from South to North, etc.," could have an impact on the American racial landscape. One committee member, considering this matter urgent, believed that black and white migration from the rural South to the urban South and North would "tend to increase race mixture. The more you get the blacks diluted among the whites, the more there will be this crossing going on."[56]

One of the most interesting passages from the conference transcript comes from Charles Davenport, who, in the first decade of the twentieth century, had published several books and papers on the subject of race and went so far as to suggest that "as the term is employed by geneticists," racial groups "may be said to belong to different elementary species."[57] Yet at the 1923 conference meeting, Davenport instead argued that "pure races do not occur in human beings," and even suggested an alternative definition—a segregate group—for some human populations. A segregate group, rather than a racial group, would have experienced "more outcrossing" and "less endogamy," which, in turn, makes it a "less pure race."[58]

Davenport characterized the populations of central Europe, for example, as largely a segregate group rather than a racial one. And although Davenport did not believe that in the wake of immigration "races are going to become altogether extinct and that all mankind will become mixed in one homogenous group," his ideas of segregate populations seemed to anticipate the emergence of the idea of ethnicity as a category that would gradually supersede race for European immigrants to the United States.[59]

As far as the biological effects of racial intermixture, Davenport advocated for investigations into physical, temperamental, and mental differences produced by racial amalgamation. He was concerned specifically about segregate groups arising from black and white mixture, noting that "we see the defects in the mixture of the Negro-white—the offspring often have the push and determination of the white but the intelligence of the Negro; they are dissatisfied with themselves and with the world; they do not have ability to better their conditions."[60] At the conference, in order to study these effects, Davenport suggested Jamaica and Bermuda as potential locales with accessible "records of white-Negro matings."[61] Davenport would spend much of the rest of the decade collecting data and developing theories about the nature of black-white interbreeding in his work on race crossing in Jamaica.[62]

Raymond Pearl, the prominent Johns Hopkins School of Public Health biostatistician and geneticist, was a major beneficiary of the Committee on Scientific Problems of Human Migration's generosity. Pearl's studies of what he called "racial pathology (as indicated in colored and white) from the autopsy and clinical records of the Johns Hopkins Hospital" were funded by the committee. The committee funded Pearl's study for a year in the amount of $6,000 beginning in July 1923.[63] Seeking to investigate links between the "nature, degree, and etiology of lesions, weight of organs, clinical history, etc., in relation to race, especially European and American stocks, and the Negro," Pearl's research program produced eight papers on three topics related to race pathology: race and alcoholism, race and cancer, and race and tuberculosis.[64]

Pearl was an aberration in the eugenics movement: he was one of the first dues-paying members of the Eugenics Research Association, a member of Davenport's inner circle, and his work provided more mainstream "evidence" for scientific racism. Yet Pearl sometimes publicly opposed racism, worked with NAACP president Walter White, and, at times, opposed the eugenics movement itself.[65] Among Pearl's papers are two sets of surviving correspondence with White: a friendly exchange in 1933 in which White asked Pearl to serve on a committee studying Harlem Hospital, and a letter to White from Pearl relaying the results of a study he was about to present at the American Association of Physical Anthropologists showing that "innate fertility" between blacks and whites "appears to be identical."[66] Pearl's opposition to eugenics was rooted not in its core canon but in what he believed was its sometimes inappropriate scientific claims. Thus, his opposition to eugenics as science could exist alongside his own personal racism and anti-Semitism. Pearl believed that eugenics was a social movement, but believed genetics to be an academic discipline.[67] This would account for Pearl's pressure on the committee to develop better methods to study human diversity.

Rankled by the methods and conclusions of some eugenics research as early as 1922, Pearl dismissed Charles Davenport's research on the matter of racial hybridization. In November 1922, Pearl responded to a letter from Yerkes seeking his analysis of a paper published by Davenport in 1917 titled "The Effects of Race Intermingling." Yerkes asked Pearl whether he thought Davenport's conclusion in this paper—that racial hybrids result in "bad hereditary combinations"—was accurate.[68] "I think the statement of Davenport's which you quote," countered Pearl, "actually rests upon very little except *a priori* reasoning."[69] In a follow-up letter, Yerkes notes that he "had the feeling as I read Davenport's article that he was expressing wishes rather than facts."[70] Pearl's very public break with the eugenics movement came in 1927. In an article published in H. L. Mencken's magazine *American Mercury*, Pearl attacked who he believed to be the scientific charlatans who had filled the study of genetics and human differences with "emotional appeals to class and race prejudices, solemnly put forth as science, and unfortunately accepted as such by the general public." Pearl also bemoaned the propaganda and scientific "phases" of the study of human traits as "almost inextricably confused so that the literature of eugenics has largely become a mingled mess of ill-grounded and uncritical" study.[71]

Yet despite his role as the curmudgeon of the eugenics movement, his sometimes public opposition to the class and racial presumptions of eugenics, and his insistence on improving the methods of eugenic study, Pearl's work would strengthen and further the cause of racial science. In the late 1920s, for example, Pearl published a series of seminal articles in *The Archives of Pathology and Laboratory Medicine* about pathology and race difference. In those articles on necropsy studies on cancer patients in Baltimore County, Pearl's data shows that cancer rates are higher in whites compared with blacks and that "the susceptibility to malignancy of the different organ systems of the body is a more diffuse and widely scattered phenomenon in the white race than it is in the colored."[72] From this data Pearl concluded that "malignancy is, in certain respects, a different condition in the white race from what it is in the colored race," thus reinforcing the popular notion that black and white biology were fundamentally different.[73] Pearl did not always think this way. Just a year earlier, in 1927, Pearl believed it "possible that these differences in the biologic and physical environment play a significant part in determining the observed differences in cancer incidence in these two broad racial groups. The main point to be kept in mind for the present is that the evidence so far reviewed permits no final conclusion as to whether the greater susceptibility of the whites than of the colored to cancer finds its origin in inherited racial differences, or in purely environmental differences."[74]

His data on the "site of primary carcinomas" show that "what cancer colored persons have tends to appear primarily either in the alimentary tract or in the reproductive organs, and not in the other organ systems, more constantly and regularly than is the case with white people."[75] This led Pearl to a curious conclusion: "White people are further along on the evolutionary pathway in respect to cancer than the Negro."[76] To come to this conclusion Pearl would have had to believe the following: cancers of the alimentary tract and reproductive organs (more prevalent in blacks) were somehow more primitive, and, therefore, the sites of cancers in whites somehow represented an evolutionary advance. In a paper called "Evolution and Mortality," also published in 1928, Pearl compared "biologically classifiable deaths" over four years between humans in London, England, and São Paulo, Brazil, and between reptiles, birds, and mammals at the London Zoological Garden. Here he found confirmation of his cancer-evolutionary hypothesis.[77] Regardless of whether the data showed an increase or a decrease in death rate by organ system between lower and higher order of animals, Pearl concluded that this represented an evolutionary advancement. And he placed this in a racial context by "showing" that the causes of death by organ system in men in São Paulo (whom he calls a more primitive population, "from both a general evolutionary viewpoint and from that of public health and sanitation, than that of England and Wales") fit nicely between the men of London ("a high product of human evolution") and lower order animals.[78]

Given Pearl's membership on the committee, it should come as no surprise that careful attention to methods was part of the group's work. An important item from the proceedings of the committee was its focus on the need for "improvement in method, devising and perfecting tools for the solution of the problems" related to migration and race. In its various proceedings the committee recommended careful attention to investigating methodological issues, primarily those related to deficiencies in the biological and psychological study of race and migration. At the committee's 1923 conference, Pearl spoke of the importance of refining methods of studying race, stating that "whatever problem we undertake, or the committee undertakes, must be studied either with already existing materials of a statistical character, or else it must be done with materials yet to be collected. It is not easy to get really critical material on these problems."[79]

Of the sixteen projects funded by the committee between 1923 and 1926, almost all of them sought to improve the epistemological approach to the study of race. Among the committee's priorities in this area were the consideration of what qualitative characters were needed in the study of physical inheritance, and what race stocks would be "readily accessible" for experimentation. Also, the committee hoped to develop methods to better understand "human inheritance through the study of family strains."[80] Attention to psychometric methods also furthered racial studies. Grantees sought to develop an "effective means of measuring mental, motor, and emotional traits." The application of this information to race crossing was considered a priority. Among the studies funded supporting these lines of research were Joseph Peterson's "Comparative Psychological Study of Negroes and Whites" (funded between 1924 and 1927) and Carl Brigham's "Internationalizing or Universalizing Mental Measurements" (funded between 1923 and 1926). Finally, Clark Wissler's "Behavior of

Physical Traits in Race Intermixture" (funded between 1923 and 1926) was designed to perfect a technique for the "study of human physical traits in pure and mixed groups and to measure the physical characteristics of Negroes, Indians, Japanese, Hawaiians, Whites, and crosses."[81]

That committee members sought to develop methodologically sound studies for "estimating the effects of race intermixture" in an attempt to distinguish their brand of racial science from years of work by eugenicists in the same area seems a moot point given the committee's makeup; nearly all committee members were eugenicists in some form, and Davenport was the titular head of the movement. But their desire to develop scientifically rigorous methods was more than just political and scientific expedience (though this was certainly part of their calculus). It would seem that committee members were genuinely interested in refining the biological and psychological study of race. We see this in the nature of the studies funded by the committee and the papers published as a result of this research, and we see this in the discussions, correspondence, and reports of the committee. Ultimately, the focus on methodology illustrates the ways in which these NRC committees were part of the scientific reimagination of race differences during this time.

In addition to shaping debates on race and immigration, the work of these NRC committees on race matters during the 1920s also had a lasting effect on the methods utilized in some disciplines of the social and natural sciences. Raymond Pearl's work, for example, was part of a larger trend in demography that sought to develop a more robust understanding of the changing nature of populations.[82] Additionally, the work of Clark Wissler and other committee members had a marked impact on the field of human intelligence testing.[83] In his final report of the Committee on Scientific Problems of Human Migration, Wissler outlined what he considered the many achievements of the committee's research program: "a series of tests to minimize language handicaps in mental measurements; a special group of tests for rating and analyzing mechanical aptitude; some fundamental pioneering in the analysis of personality . . . ; an attempt to reach the fundamental psycho-neural responses . . . ; an effort to develop an approach to organic differences in peoples through data as to pathology; and an attempt to test out qualitative anthropometric characters as a method in the analysis of mixed races."[84]

During the second half of the 1920s two new committees took over the mantle of racial research at the NRC—the Committee on the Study of the American Negro and the Committee on Racial Problems. In these new committees, the NRC shifted resources toward what it called the study of the Negro. This should not, however, belie the fact that the work of both the Committee on Race Characters and the Committee on Scientific Problems of Human Migration had already garnered significant NRC resources toward the study of African Americans. The work of these two committees provided the groundwork for some of the most influential work on race during the 1920s and for the work of the subsequent Committee on the Study of the American Negro and the Committee on Racial Problems. Indeed, research focused specifically on the study of African Americans and on race in a black-white context reflected the NRC's ongoing commitment to race.

# 5

# COLORING RACE DIFFERENCE

B y the middle of the 1920s, the National Research Council's (NRC) race committees had transitioned from the study of white ethnics to the study of black-white differences, resulting in the formation of the Committee on the Study of the American Negro. In the wake of the anti-immigrant Johnson-Reed Act of 1924, which closed the door to new immigration, and growing concern over how a demographically changing African American population would impact American society, the NRC felt a need to understand the "vital statistics of the Negro population." Hence, it diverted the focus of its racial research to concentrate on issues related to American blacks. This new committee believed its scope of inquiry should "extend over all the biological and psychological aspects of the life of the full-blood Negro and of the mixed population."[1]

The genesis of the NRC Committee on the Study of the American Negro seems to have been an October 1925 letter to G. M. Stratton, then the chairman of the Division of Anthropology and Psychology, from Robert Terry, a professor of anatomy at the Washington University School of Medicine in St. Louis. Terry was a participant in several NRC race-related committees during the 1920s. As a physical anthropologist of some renown, he contributed to the debates on race through what he believed to be a "problem fundamental to the studies of the colored population of the United States: the physical constitution of the American Negro."[2] In his letter, Terry judges that "there is no more serious work to undertake than the study of the colored races under our government," and that "there will be everlasting regret if the problems of the colored races are not taken up seriously and vigorously now." Terry, fearful that miscegenation would somehow deprive researchers of an understanding of the "colored races," hoped that the NRC would initiate research in the areas of "vital statistics of the Negro population," "the reproductive period in the American Negro woman," and the "study of somatological variation" among American blacks.[3]

A January 1926 letter to Stratton from A. E. Jenks outlined a genetically minded rationale for shifting resources to the study of black Americans. In his letter, Jenks expressed "hearty agreement with Dr. Terry's suggestion for study of the Negro." Jenks offered three reasons that reflected the continued impact of eugenic views of human difference on America's racial debates: first, he suggested "there is the general opinion that such persons [African Americans] are less fit than the remainder of the nation" and that this was a problem unique to America (excepting "the newer governments in South Africa"); second, the populations of what he considered both pure and mixed blacks were in sufficient numbers to "secure the most convincing sets of data"; and, third, "while all our so-called racial groups should be studied, the 'Negro' is evidently hereditarily less like the large numbers of white groups than those white groups are like one another, so a comprehensive study of the Negro would probably contribute distinctly toward a knowledge of so-called hereditary race characters more than would an equally comprehensive study of some one white group."[4]

Yet despite the eugenic influences behind the formation of the committee, the membership of the group reflected a diversity of opinions on matters of race. After all, antiracist and antieugenicist Franz Boas and the racist and eugenicist Charles Davenport were both members of the committee. Committee members also included E. A. Hooton, Aleš Hrdlička, T. Wingate Todd, and R. S. Woodworth.[5] Furthermore, the committee sought to complement its work on this subject in the natural sciences with the expertise of the Social Science Research Council (SSRC). Correspondence from the committee recognized the SSRC's expertise as "indispensible," "since the biological characteristics of man do not depend solely upon descent but also upon social environment."[6] However, despite repeated attempts to nurture a collaboration between the two groups, no formal partnership was successfully developed between the NRC Committee on the Study of the American Negro and the SSRC.[7]

While the stated goals of the committee included "acting as a fact-finding, stimulative, and coordinating agency in the field of Negro research," it seems as if it accomplished little during its three-year existence.[8] This failure seems due, in large part, to a "lack of funds for the prosecution of research."[9] Only research that did not require financial support was undertaken by the committee, such as lists of universities and investigators that had research in progress on African Americans; a bibliography on research on African Americans; and the compilation of already existing vital statistics on African Americans.[10]

The one considerable success of this committee's almost three years of work was the "Conference on Racial Differences" held at the National Academy in Washington, D.C., in February of 1928 under the joint auspices of the NRC and the Committee on Problems and Policies of the Social Science Research Council. The conference, "called to consider the coordination and facilitation of research on problems of racial differences and racial changes," focused its efforts on "the Negro and the Immigrant in relation to whites and stocks of earlier introduction." It was a "who's who" of the natural and social scientists of the time.[11] Among the twenty-five attendees were distinguished academics including Franz Boas, E. A. Hooten, Fay-Cooper Cole, Raymond Pearl, and Melville Herskovits; eugenicists such as Charles Davenport, Robert Terry, and T. Wingate Todd; and representatives from foundations such as W. W. Alexander from the SSRC, Leonard Outhwaite from the Laura Spelman Rockefeller Memorial, and Graham Taylor Jr. from the Commonwealth Fund.[12]

The conference was organized as a series of discussions held over a two-day period. Among the topics discussed were methodological challenges to measuring intelligence, disease, pathology, and anatomy among and between racial groups; the scientific meaning, if any, of "race" itself; the characteristics of the "African" body type; the potential dangers of racial hybridization; the sociological nature of black culture; and the origin of racial attitudes and prejudices.[13] The proceedings were generally cordial, although it is possible that the transcript of the event deadened striking disagreements between participants on a host of subjects. When disagreement did arise, discussion always seemed circumspect and respectful. It was indeed an academic affair. Yet, reading through the proceedings of the conference offers a unique window into how men of science were

struggling to understand and shape a common definition of human difference and how, ultimately, this struggle helped to endow the term "race"
with biological and genetic meaning.

Fay-Cooper Cole, an anthropologist at the University of Chicago,
opened the proceedings with a discussion about the difficulties in studying and defining the term "race." Through his work on Asian Pacific cultures (in Malaysia, the Philippines, and Indonesia), Cole was well aware of
the methodological and practical challenges of using the term "race," and
he shared these frustrations with the scholars gathered at the conference.[14]
Given Cole's wariness of the term "race" itself, it is interesting that he was
chosen to open the proceedings, although no correspondence could be
found that indicates that this was intentional one way or the other. Cole
was particularly frustrated by two things: the fact that the term "race"
was "frequently used in three or four ways in the same article" and that
"while there is a great deal of confusion in the use of the word, appar-
ently the general public thinks that those of us who are working in this
field know what we mean by the term."[15] Cole's candidness on this matter
underscores the contradictions of even the most vehement proselytizers
of the term and the word's biological meaning: that in private discussions
among colleagues some scientists could admit to their misgivings about
the idea of race but in public, be it in scientific journals or popular print,
the meaning of race seemed unambiguous. It is surprising that Charles
Davenport, whose public statements were among the most influential in
garnering support for a biological and genetic view of race during the
second and third decades of the twentieth century, could be circumspect
in private about the meaning of human difference.

Boas, the third speaker of the conference, spoke about "Changes in
Immigrants," a presentation that drew upon his well-known work in the
area and concluded that "American-born descendants of immigrants
differ in type from their foreign-born parents."[16] He described his findings that anthropometric measurements such as head form, stature, and
facial width, once believed to be fixed by heredity, could instead vary
by environment.[17] These findings were revolutionary in anthropology
and would come to have a significant impact on racial thinking more
generally. By showing that the environment mediated intergenerational
difference among immigrants, Boas's work highlighted the importance

of environmental factors in determining physical traits and challenged scientific theories about the permanence of racial traits.[18]

Boas argued in his talk that heredity was not solely responsible for an individual's body type or phenotype. Instead, he believed that evidence showed that body type was influenced by both heredity *and* environment. Boas, also wary of the methodological efficacy of "race" (like Fay-Cooper Cole), warned of the shortcomings of using race to study heredity in human populations. Boas reminded his fellow attendees that "race, as we use the term, is composed of a great many different family lines," and that in order to compare "a race in one environment with the same race in another social or geographical setting we must be certain that the same family lines are represented in both series." In other words, "races" were not useful to the study of human heredity because racial groups are made up of diverse "family lines" or subpopulations.[19] Boas was one of the first scientists to recognize that skin color, despite being the definitive trait for a race, was just one of many traits that defined populations and thus made using racial groups highly problematic in the study of complex patterns of heredity and the relationship between heredity and environment.

Boas also articulated what would soon become recognized among population geneticists as a fatal flaw in applying the race concept to the study of human heredity: that "family lines in all so-called 'races' may be much more different among themselves than family lines that happen to belong to two different races."[20] That Boas understood that populations within so-called races could be more dissimilar than populations between so-called races was an important intellectual jump in the long argument against racial thought, and Boas was among the first of a small group of scientists who began to apply this thinking to race in the 1920s and 1930s. Even as Davenport agreed with the spirit of Boas's misgivings about race, he remained opposed to the specifics of the new approach to human diversity as articulated by Boas. In Davenport's comments to Boas, he worried that the study of heredity was "even more complicated than he [Boas] has stressed," and that "part of the difficulty, I suppose, lies in the definition of race or type."[21] Boas acknowledged his and Davenport's (as well as many of the other conference attendees') shared frustrations with "race," saying that "we both, I believe, wish to discard the term 'race' and lay stress upon the family lines."[22] There is no recorded response by Davenport to this assertion in the conference proceedings.

Even with his reservations about the definition of race, ultimately Davenport believed in the concept, however muddled his approach. The differences between him and Boas, still to this day, generally define the lines along which racial and antiracial thinking persist: population versus typological thinking. The biologist Ernst Mayr frames the differences between the two men in this way: "All groupings of living organisms (including mankind) are populations that consist of uniquely different individuals. . . . Populations differ from each other not by their essences but only by the mean differences of statistical populations. Population thinking is an entirely different way of thinking from typological or essentialistic thinking." Mayr tells us that for typologists, or essentialists, variety "consisted of a limited number of natural kinds (essences, types), each one forming a class, the components of which are essentially identical, constant, and sharply separated from the components of other such essences." Variation by this way of thinking is "non-essential and accidental." That is why a typologist looks at skin color and sees not only, for example, whiteness but also all the traits thought to be associated with that color. When it comes to differences in all species, a typologist does "not know how to deal with variation. This is particularly conspicuous in his treatment of the human races. For him, whites, blacks, Asians and so forth are types that invariably have certain racial characteristics."[23] This is why in one breath Charles Davenport could agree with Boas that the definition of "race or type" caused biologists, geneticists, and anthropologists methodological difficulties, but in his next breath Davenport could ruminate on the specific effects racial hybridization would have on resulting generations.[24]

It is fair to say that Mayr's creation of the antipodes of population versus typological thought had a much more complex and nuanced dynamic within the evolutionary sciences, including population genetics. Discussions about variation dating back to Darwin were central to arguments about evolution, and central to the emergence of the evolutionary synthesis in the 1930s.[25] However, this does not contradict the way fairly fixed ideas about typology undergirded arguments about variation, particularly in humans, within the biological sciences during the first three decades of the twentieth century. It wasn't really until Theodosius Dobzhansky's work on the evolutionary synthesis that ideas about variation in human populations would become more integrated into a population genetics worldview.

Davenport's absolute adherence to typology without an understanding or application of population genetics to his work was quickly making him obsolete as a geneticist (although he managed to retain prominence as a eugenicist and some distinction as a geneticist into the early 1930s). Around him his field was quickly changing—his failure to completely embrace the chromosomal theory of heredity certainly did not help matters—as eugenical views of human and other organismal diversity were gradually being shown to be obsolete by work in the area of population genetics. During the 1920s geneticists Ronald A. Fisher, Sewell Wright, and J. B. S. Haldane, among others, developed the mathematic and theoretical foundations for the synthesis of Mendelism and Darwinism, and helped make a populationist view of organismal diversity possible.[26]

However, even as geneticists and other scientists provided evidence that undermined racial typology, there were still those among this new breed who held fast to typological views of race. For example, R. A. Fisher's *The Genetical Theory of Natural Selection* at once provided mathematic models undermining typology *and* dedicated several chapters in this now classic genetics text to eugenics. Fisher's eugenic sentiments are unambiguous. "The most obvious requirement for a society capable of making evolutionary progress," he wrote in his classic text, "is that reproduction should be somewhat more active among its more successful, than among its less successful members." That Fisher could at once be the genius behind many of the mathematical models of modern population genetics and a eugenicist is in many ways similar to Raymond Pearl's contradictory relationship with eugenics. Fisher was sympathetic to the ideas and worldviews of the eugenicists, he just thought that the methodology underlying their theory was wrong. His daughter and biographer explained Fisher this way: he "could never let pass what he believed to be wrong reasoning. He was known on other occasions to argue the case, even with someone who produced a correct answer, if the reasoning on which it was based was not sound."[27] Likewise, Edward

East, the pioneering population biologist whose experimental work showed how Mendelian inheritance "could account for an almost continuous array of variation," also authored the racist tome *Mankind at the Crossroads*. In the chapter "Racial Prospects and Racial Dangers," East wrote in typological terms about the fixity of differences between black and white populations: "The Negro as a whole is possessed of undesirable transmissible qualities both physical and mental, which seem to justify not only a line but a wide gulf to be fixed permanently between it and the white race."[28] According to the geneticist Bentley Glass, "No one went farther than East in lending his authority to racist and social prejudices."[29] While this point is debatable—Davenport, after all, was also a classically trained geneticist who was a product of the same Harvard institute where East taught—East certainly was among the most prominent racists among the generation of geneticists who were helping to develop population genetics.

The contradictions and tensions between population thinkers and typologists were in evidence at the "Conference on Racial Differences," and the conference may be unique as the first forum where these opposing viewpoints were debated and divergent views on race were shared in such an intimate way. What seems so remarkable about the conference is that it seemed, intentionally or not, to set the parameters for the debate about the biological meaning of race for the remainder of the twentieth century. And not just in the simplistic population-versus-typological approach that so dominated the conference. What still seems striking today is that despite almost eighty years of science, the debate has changed so little.

In addition to Fay-Cooper Cole and Franz Boas, the populationist position was promoted by the anthropologist Melville Herskovits, a Boas acolyte and young assistant professor at Northwestern University, who spoke about "The Role of Social Selection in the Establishment of Physical Type." In describing the emerging opposition to the race concept, Herskovits said that "the existence of a group of such heterogeneous descent and termed 'Negro' is due to the fact that we have a social definition of the word Negro, and we do not have the same meaning in mind when we use the word as the anthropologist who does when he speaks of the African Negro."[30] Herskovits also insisted that the social as well as the biological side of individuals must be recognized and studied.[31] Yet despite Herskovits's place in the populationist camp, his conclusions were somewhat contradictory and illustrate how a populationist thinker could also embrace typological arguments. On the one hand, for example, Herskovits argued in *The American Negro: A Study in Race Crossing* (published the same year as the conference) that race is a concept used "with amazing looseness" and recognized "how little we are able to define a word that has played such an important role in our political and social life." But Herskovits also believed that his data showed that the American Negro was "a homogeneous population group, more or less consciously consolidating and stabilizing" due to both legal (antimiscegenation statutes) and social (segregation and racism) pressures. Herskovits did not, however, take the step that a pure typologist would have: he specifically did not regard the American Negro as a new race, no matter that his data showed that the population was homogeneous. Indeed, at the end of *The American Negro* Herskovits warned that to do so would be fallacious, and that such thinking, "translated into action in the field of race, too often makes for tragedy."[32] Part of Herskovits's limitation was the nature of his methodology—by relying on self-reported genealogies his data could show the emergence of a homogeneous population. By allowing his study subjects to self-identify their race, Herskovits probably hoped to avoid the gross typological observations of some of his colleagues. Yet in doing so he was relying upon the observations and memories of individuals who were still a part of the American system of racialized thinking, even though they themselves could be its worst victims.

On the other side of the hereditarian spectrum were speakers like Raymond Pearl, whose talk on the "Incidence of Disease According to Race" argued that the "statistical characteristics of disease do have a rather definite correlation with race."[33] So too did T. Wingate Todd expound on hereditarianism. Todd, an anatomist at Western Reserve University (later Case Western Reserve), spoke on "The Search for Specific African Body Features." Todd's research, blatantly typological in nature, argued that "the real distinction between our Whites and our Negroes is, then, in absolute dimensions." And even as he used measurements to "subdivide our Negroes according to their white characteristics," he also believed that "the American Negro is becoming homogeneous."[34]

Finally, the psychologist Joseph Peterson presented on "The Problems and Results of Negro Intelligence." Peterson, a psychologist at George Peabody College for Teachers in Nashville (now part of Vanderbilt University), was, perhaps, the South's most distinguished psychologist at that time, and he wrote extensively on issues pertaining to racial intelligence. At the time of his death in 1935 Peterson had even risen to become president of the American Psychological Association, the first professor from a southern university to hold that distinction.[35] His 1923 book *The Comparative Abilities of White and Negro Children* studied over 3,000 white and black children from the South and concluded that white children performed better on intelligence tests.[36] Yet Peterson was hesitant to conclude that such results overtly indicated African American inferiority, and he also questioned the accuracy of the tests, writing that "work in this line has been premature and untrustworthy; nevertheless many testers have been ready with conclusions." Moreover, Peterson was less concerned with this question than with using intelligence testing for what he called better adjustment, which was, in a sense, a "gentleman's" way of articulating racism. Hoping to use testing to serve a beneficial social purpose, Peterson sought to use test results mainly for the "sort of self-control that the sound use of tests of any kind engenders." In other words, he believed testing could lead to the "voluntary regulation of birth rate and elimination by eugenic methods within each race (or national) group of undesirable physical or mental traits, stronger in certain individuals than in others."[37]

Peterson's views were not an anomaly in the field of intelligence testing. A 1934 survey of seventy-seven psychologists, thirty education scholars, and twenty-two sociologists and cultural anthropologists on the subject of race and intelligence included as subjects Peterson and other well-known leaders in the field like Charles Davenport (who is oddly classified as a psychologist), Jean Piaget, Carl Seashore, John Dewey, Franz Boas, Fay-Cooper Cole, E. A. Hooton, and Knight Dunlap. The survey found that most of those in the fields of education and psychology believed "race inferiority possible but not adequately demonstrated," whereas most anthropologists and sociologists (65 percent) were "highly critical of the means used to demonstrate race inferiority and of the results obtained."[38]

At the conference, Peterson raised several questions that would vex the field of IQ studies for much of the twentieth century. Peterson, for example, suggested that "to determine the intelligence of Negroes is a rather difficult problem" because "we do not know exactly . . . what the Negro race is" and we "do not know what intelligence is." He proposed, as he did in his books and published articles, that the intelligence testing that had so far been done on black Americans had "been of a casual nature": "Someone happens to find it convenient in his academic position to do some Negro testing and goes out and gives tests that are used for whites. He gets certain data and compares those data of Negroes in one community with the norms of whites, and puts down the difference as a race difference." Yet for all his skepticism about the nature of intelligence testing and of racial differences in intelligence, in the end Peterson's perspective would help to shape the racialized IQ debates of the twentieth century. He might have questioned the scientific legitimacy of racial categories and the definition of intelligence, but he still came back to the question, "Is the great retardation of the Negroes due entirely to lack of opportunity or is it partly at least due to innate deficiencies?"[39] And it was Peterson's attempted solution to this question that was the most troubling actionable item to come out of the conference.

In the hope of developing a new methodology that could settle the question of innate differences in group intelligence, Peterson proposed an experiment whereby "complete control of the group of individuals selected as representatives of the races to be compared, complete control through a long period of years, including their entire school training. . . . If we could get an experiment like that going, and give the children of the compared races essentially the same sort of training throughout, then we should be able to throw some light on the question as to the degree of innateness of any differences found."[40] Although there was little discussion of this item at the conference itself, Peterson's proposed racial or IQ orphanage was among the research projects recommended for further study by the group. This project would be picked up by the Committee on Racial Problems, jointly formed in March 1928 by the NRC and SSRC to formally act on the conference's recommendations.[41]

The conference closed with several proposals for "research on the scientific problems of race." In addition to Peterson's experiment, the conference sought support for a five-year study of comparative race pathology to be carried out by Raymond Pearl; Robert Terry's proposal to utilize

cadavers from the nation's medical schools to address the need for anatomical data on race difference; Aleš Hrdlička's proposal for a study of
the physical anthropology of immigrants; a methodology to study the
impact of heredity versus environment "to be applied to immigrants as
well as Negroes"; a study examining the physical and cultural components
of populations in the Americas, Caribbean, and Africa; and a "proposal
for a study of 'race-crossing' in a large city and nearby rural community."[42]

However, surviving records suggest that the Committee on Racial
Problems focused its efforts almost exclusively on Peterson's idea of creating racial orphanages to measure biological and environmental differences between black and white races. This may be partially explained by
the fact that Knight Dunlap, chairman of the Division of Anthropology
and Psychology, referred one of the proposals from the conference—
Raymond Pearl's project "looking toward a 5-year study of comparative
race pathology"—to the NRC Division of Medical Sciences. The remainder of the proposals were, according to minutes from the committee's
meeting on January 12, 1929, tabled, deemed unactionable, or recommended for further study.[43] The final report that the committee issued in
April 1931 makes no mention of other projects resulting from the recommendations of the "Conference on Racial Differences."[44]

The efforts of the committee went solely toward the development of a
second conference, May 25, 1930, that would explore the "feasibility of a
plan for studying the comparative development of children of different
races in the hope of throwing light on the contributions of heredity and
environment in determining racial differences."[45] For reasons that remain
unclear, Joseph Peterson, who proposed the racial orphanages, was not
involved in the Committee on Racial Problems, nor in the 1930 Detroit
conference. Instead, representing the Joint Committee on Racial Problems
were the neuropsychologist Knight Dunlap, of Johns Hopkins, American
Museum of Natural History anthropologist Clark Wissler, Columbia University psychologist R. S. Woodworth, Cornell University psychologist
Madison Bentley, and sociologist Robert S. Lynd from the SSRC.[46] That
the NRC thought that such a plan could be put into action reflected not
only the prevailing racial views of the time but also, to a degree, the Progressive Movement's concern for child welfare and social control. Under
the Progressive banner, intervention in family life to protect a child was

considered normal and necessary. However, by the 1930s, as the Progressive era gave way to the Great Depression and the New Deal, support for the construction of orphanages, let alone the proposed racial orphanages, was probably hard to come by. The era of orphanages would quickly come to a close with the rise of the New Deal welfare state.[47]

Prior to the Detroit conference both Fay-Cooper Cole and Franz Boas offered their thoughts to Dunlap on the idea of a racial orphanage. Cole was very critical, writing that the "proposal is very interesting and one which would certainly yield valuable results, but it is not without serious objections." Cole worried that the differences of care between white and black institutions could not be controlled for and would result in "differences in experiences." He believed that this was because it would be impossible to "alter the experiences of the wardens and nurses so that they would react equally to black and white skins."[48] Surprisingly, Franz Boas deemed the experiment "highly desirable." He hoped "the committee in charge of this matter which has been in touch with the Negro problem and knows by personal experience the prospects and difficulties" would see it through. Boas also had hoped for a "Negro on this committee."[49] In the end, neither of Boas's suggestions seems to have been heeded.

The proceedings of the Detroit conference explored the methodological challenges of developing an experiment to "throw light on the influences of heredity and environment in producing the differences actually observed in present-day groups in health, vigor, mental achievements and social adjustments." "Plan A" called for the development of an institution that would "receive" children at birth or in early childhood to provide "an environment superior of that of the private home." In this controlled environment, researchers would be provided with an "opportunity for intensive study of development." "Plan B" would see the construction of nursery schools "with influence brought to bear also upon the homes from which children were received." It was hoped that in such a setting, the general environment of the student could be drastically improved, and also allow for "following development." Finally, "Plan C" would follow black children placed in "superior negro foster homes, with special care to secure an adequate control group."[50] The group considered the challenges to all three plans, including the barriers to securing subjects; the barriers to securing orphanage staff; and the sampling problems faced by such an experiment.[51] Ultimately, they worried about social and cultural biases, acknowledging that "a complete control of social and cultural factors would be pretty difficult. . . . A single accident, such as the introduction of cultural bias, might do a great deal to vitiate the experiment."[52] In the end, the idea for such an institution died quietly—the SSRC would soon after drop the matter and withdraw itself from the joint committee. The NRC initially appointed a committee to follow up on the idea of a racial orphanage, but that committee never took action on the idea.[53]

Even though the racial orphanage experiment never moved beyond the planning stage, it nonetheless raises troubling issues about the intersection of race and scientific experimentation, and the role that federally funded science played in shaping the race concept. The proposed orphanage study also illustrates how racial ideology shaped research agendas, research protocols, and, of course, the interpretation of research data, much as it did for "The Tuskegee Study of Untreated Syphilis in the Negro Male," a forty-year experiment examining the natural history of the disease in 399 untreated syphilitic African Americans living in Macon County, Alabama.[54] Both studies epitomize the way eugenic thinking about race corrupted those seeking to understand and address scientific questions and medical challenges.

With the dissolution of the NRC Committee on Racial Problems, the NRC's careful attention to race difference, for the time being, came to a close. During the 1920s the four NRC committees on issues of race exemplified the changing scientific approach to race in the United States. First, as the committee's work illustrates, there was a dramatic shift in how racial groups were defined during this period, from a broad definition that included both continental ancestry as correlated to skin color and ethnic, national, and religious affiliations to one that focused primarily on skin color. Second, the decade saw the emergence of large-scale research projects into the biological nature of racial differences between white and black Americans. These studies focused primarily on intellectual, pathological, and morphological differences between the two groups. Third, research sponsored by the committees sought to understand the biology of the children of black and white couples, or racial hybrids, as they were often called. And, finally, the decade witnessed the emergence

of populationist thinking as an alternative to typological thinking, specifically as it concerned race difference.

## DAVENPORT'S RACE CROSSING

In many ways the 1929 publication of Charles Davenport and Morris Steggerda's *Race Crossing in Jamaica*, a study examining the biology of black-
white racial hybrids in Jamaica, embodied the state of racial research,
particularly the research generated by the NRC during the 1920s. The
work, after all, was proposed in the early 1920s to the Committee on
Human Migrations of the NRC. The book was a field study seeking to
measure intellectual, pathological, and morphological differences between
blacks and whites, as well as to understand these traits in hybrids between
these two groups. The study remained wedded to a typological approach to
race difference, yet as critics of the work would point out, the data collected
did not necessarily correlate with its typological conclusions. Davenport's
influence on the NRC was ongoing throughout the 1920s. Though the
publication of *Race Crossing* was not directly funded with NRC dollars, its
ideas and methodologies both influenced and were influenced by Davenport's involvement with NRC committees on race, especially the work of
the Committee on the Study of the American Negro, the "Conference on
Racial Differences," and the Committee on Racial Problems. When asked
to serve on the Committee on the Study of the American Negro in 1926,
Davenport accepted, responding to the invitation with enthusiasm: "I may
say that I am vigorous in this matter and engaged in studies of it and shall
be very glad to serve on the committee."[55] In continuing correspondence
with the committee's head, Davenport highlighted and encouraged the
need for "the invention of new methods of quantitative investigation to
study this issue."[56] In 1927 he would write to R. J. Terry, urging that the
Committee on the Study of the American Negro "seek funds to carry on
investigation of the Negroes and Negro-white hybrids in one or more cen-
ters in some part of the country where both groups can be found in fairly
large numbers."[57] Terry, chairman of the committee, reached out for Davenport's assistance several times, including once in November 1928 when
Terry asked all committee members for help "in getting aid for research

on the biology of the Negro."[58] In March 1929 Davenport again wrote to
Terry, concerning the recommendations of the "Conference on Racial Differences," making suggestions for the racial orphanage study that included
involving Boas—"the chief skeptic"—in the process and making sure that
the comparison of white and black children "be made on the two groups
under as nearly similar conditions of training and culture as possible."[59]

*Race Crossing in Jamaica*, the culmination of more than two decades
of Davenport's research in the area of white-black differences, was a
giant tome, an almost 500-page analysis of traits from stature to weight
to intelligence between whites, blacks, and browns (hybrids). The book's
research was funded by a $10,000 gift to the Carnegie Institution of
Washington by Wickliffe Draper, who would later gain notoriety as the
founder of the Pioneer Fund.[60] Davenport had been courting Draper
as early as 1923, discussing with him his interest in a "bequest for the
advancement of eugenics."[61] It was not until 1926, however, that the two
found common ground. In February 1926, Draper suggested Davenport
might do "research work on the effect of miscegenation" and dedicate
himself to the "popularization of the[se] results."[62] Draper's money would
be put to quick use, and Davenport was clearly excited about what he
thought would be its impact, calling the work "of the greatest possible
importance."[63] "The more I think over your plans and the contribution
you have made toward the study of the consequences of miscegenation
the more enthusiastic I am over the prospect of getting something of
great scientific and practical utility," Davenport told Draper. Davenport
outlined to Draper their agreement on how his funding would be used:
"A study, in the most objective and quantitative way possible, of the
inheritance of the traits of pure blooded Negroes, as found in the Western Hemisphere (specifically, probably, Jamaica and Haiti) and of white,
as found in the same places with especial reference to the inheritance of
the differential traits in mulatto offspring and the distribution of these
traits in later generations."[64]

The basis of Davenport and Draper's relationship was clear; both were
deeply concerned that "the presence of the African negro in our country
may be very fateful for its future, as its increase tends to overcrowd more
and more the country to the detriment of the white race."[65] By the fall
of 1926, Draper's money was supporting Davenport's research into race

differences. The research was carried out in Jamaica by Morris Steggerda, then a graduate student in genetics at the University of Illinois.[66]

The three theses of *Race Crossing*—first, that hybridization between blacks and whites results in "the production of an excessive number of ineffective, because disharmoniously put together, people"; second, that "a population of hybrids will be a population carrying an excessively large number of intellectually incompetent persons";[67] and, third, that whites were superior in mental capacity to both blacks and browns—confirmed the thinking of many racial scientists and validated popular racist beliefs. *Race Crossing*'s significance also lay in its breadth of analysis. As one of the largest field studies of its kind to date, it endowed racial science with legitimacy and encouraged confidence in its results so much so that Davenport wrote in its pages, "The burden of proof is placed on those who deny fundamental differences in mental capacity between Gold Coast Negroes and Europeans."[68]

Reviews of the book were mixed. Writing in the journal *Social Forces*, the eminent Smith College sociologist Frank Hankins called the book "an extremely thorough anthropometrical study" that is "undoubtedly one of the very best contributions yet made to the difficult subject of race hybridism."[69] Other reviewers were not as kind. Karl Pearson, writing in *Nature*, was especially critical of the book's methodological shortcomings, specifically its data-collection methods.[70] *Race Crossing*'s most comprehensive critique came from Harvard scientist William Castle, an early pioneer in genetics and a former student of Davenport's from his Harvard days.[71] In 1924 Castle became one of the first geneticists to attack the idea that hybrids resulted in disharmonious crossings, arguing that such an idea was not supported by science but rather was an expression of personal views. Castle came to this conclusion through his own work on rabbits, which revealed no hybrid disharmonies. His findings in nonhuman species would be supported by wide-ranging studies on humans in Hawaii, Canada, and the United States published in the late 1920s.[72]

Castle's review of *Race Crossing*, which appeared in *Science* in 1930, was a crushing review of the book, arguing that Davenport's conclusions were not supported by his own data. Castle pointed, for example, to the assertions that "the leg of the blacks is much longer than that of the whites" and that a cross between them would result in a disharmonious cross, "which would put them at a disadvantage." In scouring the book Castle could find no data at all supporting Davenport's "justification for the idea that the brown Jamaicans have dangerously disharmonious combinations of stature and leg-length." In fact, Castle pointed out, this idea contradicted the book's own data that "the reputed 'much longer' leg length of the blacks turns out to be on the average longer by five tenths of a centimeter!"[73]

Yet despite such negative reviews, the impact of *Race Crossing* and of Davenport's position on miscegenation and black-white differences went far beyond academic circles. These ideas would be enduring—something Castle himself feared in his review of the book. "We like to think of the Negro as inferior," Castle wrote, and "we like to think of Negro-white crosses as a degradation of the white race." However, "the honestly made records of Davenport and Steggerda tell a very different story about hybrid Jamaicans from that which Davenport . . . tell[s] about them in broad sweeping statements." Castle was resigned that such ideas "will be with us as the bogey men of pure-race enthusiasts for the next hundred years."[74]

That Davenport had become one of the foremost antiblack and anti-miscegenation propagandists of the eugenic era should be no surprise, having argued in a variety of forums that "there is a constitutional, hereditary, genetical basis for the difference between the two races in mental tests." He had also lectured on the "interesting tendency" in the United States of the "dominant race to apply to the hybrid the name of the subordinate or inferior race," and called for the "study of the physical, mental and temperamental traits of Negro-white hybrids to the second and third generation."[75] Davenport worked hard to popularize the conclusions of *Race Crossing*. In the magazine *The Scientific Monthly* Davenport emphasized the genetic nature of race, writing that "genetical experimentation in hybridization has revealed" inherited traits can "recur in their pristine purity in later traditions." This, of course, was a dire warning about the potential effects of racial hybridization, a fact that Davenport did not hide in the article, asking, "What is to be the consequence of this racial intermingling? "Especially we of the white race, proud of its achievement in the past, are eagerly questioning the consequences of mixing our blood with that of other races." Davenport's conclusions confirmed this fear—that while physically "there is little to choose between the three groups," in intellectual capacity hybrids, or "browns" as he calls them, suffer from a "large burden of ineffective persons who seem to be muddleheaded or incapable of collecting themselves to do the task in hand," and that a "population of hybrids will be a population carrying an excessively large number of intellectually incompetent persons."[76]

Davenport's antimiscegenation science (embodied in *Race Crossing*) not only conformed to and confirmed the thinking of the time about race mixing but was also an important scientific retort to the mounting challenges against some state antimiscegenation statutes.[77] Perhaps nowhere was this more true than in Virginia, where, with advisement from Davenport, the eugenicist and white supremacist Walter Plecker, a Virginia public health official, helped shape Virginia's antimiscegenation Racial Integrity Act of 1924. Davenport offered Plecker advice on antimiscegenation-related issues—including help calculating race mixture and suggesting a change to Virginia's inheritance law that would prevent children from black-white relationships from inheriting land—all the while studiously trying to avoid intimate involvement with "the carrying out of that law" for fear of drawing the Carnegie Institution into "the matter of the administration of law in the State of Virginia."[78]

Davenport and Draper would have likely appreciated *Race Crossing*'s role in the antimiscegenation battle nearly twenty years after its publication (although they certainly would have been disappointed with the battle's final outcome). The book and its conclusions were used by the defense in the landmark 1948 miscegenation legal challenge *Perez v. Lippold*, in which the California Supreme Court declared its state's antimiscegenation statutes unconstitutional. The case was brought by a Los Angeles couple who were refused a marriage license by the county clerk of Los Angeles because they were members of different racial groups. In court arguments, the city's defense relied on *Race Crossing* "as authority for the proposition that the progeny of mixed marriages are inferior to those of purebred marriages."[79]

Despite *Race Crossing*'s staying power in eugenic and racist circles, Wickliffe Draper was not pleased with the book's limited popular impact. Draper would, in coming decades, turn his attention toward less-established scientists and other public figures who could help him with his racist program, funding in the 1930s, for example, the distribution of white supremacist Earnest Sevier Cox's book *White America* to political figures across the country. Here Draper's funding effort bore significant fruit: Mississippi senator Theodore Bilbo pledged support to Cox's and Draper's cause of repatriating black Americans in Africa. In April of 1939, Bilbo introduced the Greater Liberia Act on the floor of the United States Senate, seeking to mandate the removal of all black Americans from the United States. Draper would eventually focus his efforts through the Pioneer Fund, set up in 1937 to assist in the education of white descendants of the original settlers of the United States, and, more significantly, to support hereditary and eugenic research. Through the fund, Draper would assist in the promotion of his racist views through scientific studies. Harry Laughlin, formerly part of Davenport's Cold Spring Harbor group, became its first president. Over the course of the twentieth century, whether it was underwriting opposition to the 1954 *Brown v. Board of Education* decision, to anti-Semitic or antiblack causes, or funding studies to "prove" the inferiority of blacks and the biological soundness of white supremacy, the Pioneer Fund played a significant role in racial science.[80]

Those who would benefit from the Pioneer Fund's largesse had an ongoing and significant impact on public debates about race, framed primarily as issues of race crossing and of black racial inferiority and white racial supremacy. However, beginning in the 1930s, eugenic racial scientists would come to have a diminishing impact on scientific debates about the race concept, particularly within genetics.[81] Two related but separate discourses would take place over the remainder of the twentieth century; one by scientists involved in the emerging evolutionary synthesis who were struggling to reimagine how biology and the natural sciences defined race, and the other by paid scientific propagandists (including those by the Pioneer Fund) and their sympathizers in biology. The latter would steadfastly ignore an emerging scientific consensus on human genetic diversity and try to resurrect typologically based theories of human difference that were increasingly undermined by work in population genetics and evolutionary biology. The remainder of this book takes up debates within the biological sciences about the nature of the race concept. Pioneer Fund propagandists played only a limited role in these discussions, reacting largely to what would fast become a new genetical theory of race difference.

# 6

# BIOLOGY AND THE PROBLEM
# OF THE COLOR LINE

I n 1906 W. E. B. Du Bois, then a young social scientist at Atlanta University, issued a forceful and elegant challenge to racial science with the release of *The Health and Physique of the Negro American*.[1] In the pages of his book, Du Bois attacked the very foundation of America's racial ideology, calling into question the legitimacy of the race concept at a time when science was being exploited in the service of racist ideas and practices, and ideas about racial difference were increasingly becoming part of natural science's lexicon. Despite the boldness of the study and its importance as an act of intellectual protest, its contemporary impact was limited. As one commentator wrote, the usefulness of the work was not "realized" by African Americans at the time, and most whites were certainly "hostile to such a study" and its conclusions.[2] Yet the book's importance should not be judged simply on what might be perceived as its immediate impact. Instead, the work serves as a milestone in antiracist thinking and scholarship.

Du Bois's early writings on race anticipated the work of Franz Boas and other Columbia University anthropologists in the first half of the twentieth century on cultural relativism and their critiques of ideologies of racial inferiority.[3] More important, however, Du Bois also anticipated the lines along which many geneticists and other natural scientists would, over the course of the twentieth century, struggle with the scientific and social meanings of race. Yet given the book's limited readership, it seems unlikely that this impact was direct; it is doubtful that Theodosius Dobzhansky or Leslie Clarence Dunn, or other natural scientists working on concepts of genetic diversity more than two decades later, were aware of this document and of Du Bois's early opposition to racial science. But in many ways *The Health and Physique of the Negro American* struggles with the central theme of Du Bois's landmark 1903 study, *The Souls of Black Folk*: "the problem of the twentieth century is the problem of the color-line."[4] And it is in the negotiation of this "problem" that *The Health and Physique of the Negro American* illustrates and predicts how the natural sciences would be put into service demarcating and defending the color line. Du Bois implicitly understood the danger of this but believed that through careful rebuttal he could reveal race for what it was: an unscientific expression of America's racial mores.

The eighteen volumes of the Atlanta University series—all but the first two overseen by Du Bois—addressed a wide array of topics in African Americana. From *The Negro Church* to *The Negro in Business* to *The Negro American Family*, Du Bois and his Atlanta colleagues explored topics in black American life, utilizing what were then cutting-edge methodologies in the social sciences. Du Bois is remembered as a "founding father of American sociology," and his classic work *The Philadelphia Negro* was pioneering in its interdisciplinary use of history, ethnography, and statistics.[5] Du Bois's motivation for such a grand project was a Progressive belief in the power of scientific knowledge; knowledge, he hoped, that if shared with the general public could have an emancipatory effect on racism's stranglehold on the American zeitgeist. In the case of *The Health and Physique of the Negro American*, the clear target was on biological thinking in the area of human difference. "The world was thinking wrong about race, because it did not know. The ultimate evil was stupidity," Du Bois would later write. His "cure" for American racism "was knowledge based on scientific evidence," and in this and much of his other early work he set out to utilize the scientific method in his fight against this American failing.[6] The "scientific evidence" of eugenicists, anthropologists, and biologists would, of course, trump Du Bois's own. Despite this, Du Bois hoped that the Atlanta studies would be a "comprehensive plan for studying a human group and if I could have carried it out as completely as I conceived it, the American Negro would have contributed to the development of social science in this country an unforgettable body of work."[7]

To help accomplish this ambitious undertaking, Du Bois reached out to academic leaders to build his argument and philanthropists to fund the work. In 1905 Du Bois corresponded with, among others, the American Museum of Natural History anthropologist Clark Wissler asking for "a list of the best works on Negro anthropology and ethnology." Wissler understood, as he wrote to Du Bois, that "the literature upon this subject is very incomplete and unsatisfactory" and recommended several books and articles, including the book *The Races of Man* by the French anthropologist Joseph Deniker.[8] In Deniker, Du Bois found an education that must have shaped, to a great degree, his thinking about the meanings of race. Deniker was not a racial essentialist, and in his book he asserted that there were twenty-nine human racial groups, proposing that ethnicity was a more useful marker of understanding human biological and cultural diversity. For example, Deniker, considered whether "real and palpable groupings" of humans are "capable of forming what zoologists call 'species,' 'subspecies,' 'varieties,' in the case of wild animals, or 'races' in the case of domestic animals. One need not be a professional anthropologist to reply negatively to this question." In Deniker's judgment, "races, or varieties . . . are by no means zoological species; they may include human beings of one or of many species, races, or varieties."[9]

In advance of the "Health and Physique" conference, Du Bois had sought funding for the project from numerous sources, including the industrialist Andrew Carnegie, who had been a funder of Booker T. Washington's Tuskegee Institute. In a long letter to Carnegie, Du Bois outlined the work of the Atlanta conferences "with a view of securing if possible your financial support for this work." Du Bois lamented the fact that "so far this work has been carried on by small voluntary contributions," and hoped that an infusion of funding would help "to enlarge its scope and improve its methods of research." There is no record of a reply from Carnegie.[10]

In late May of 1906 Atlanta University hosted the conference whose proceedings would soon be edited into *The Health and Physique of the Negro American*. Among the conference participants were Franz Boas, who titled his talk "Negro Physique," Du Bois, and R. R. Wright, of the University of Pennsylvania, who spoke on "Mortality in the Cities." Eight months before the conference, Du Bois sought to foster a collaboration with Boas, writing to him that there is "a great opportunity here for physical measurement of Negroes. We have in Atlanta over 2,000 Negro pupils and students who could be carefully measured. We have not the funds for this—has Columbia any desire to take up such work?" Boas told Du Bois that he thought that there was nothing "particularly good on the physical anthropology of the Negro"; he rejected any proposed collaboration and request for support. Nonetheless, Boas was "very glad to hear" from Du Bois that he intended "to take up investigations on this subject."[11]

Two days following the conference Boas would deliver the commencement address at Atlanta University, in which he confronted the issue of alleged black biological inferiority. "To those who stoutly maintain a material inferiority of the Negro race and who would dampen your ardor by their claims," Boas declared, "you may confidently reply that the burden of proof rests with them, that the past history of your race does not sustain their statement, but rather gives you encouragement. The physical inferiority of the Negro race, if it exists at all, is insignificant when compared to the wide range of individual variability in each race. There is no anatomical evidence available that would sustain the view that the bulk of the Negro race could not become as useful citizens as the members of any other race. That there may be slightly different hereditary traits seems plausible, but it is entirely arbitrary to assume that those of the Negro, because perhaps slightly different, must be of an inferior type." Boas's caveat—"if it exists at all," referring to the "physical inferiority of the Negro race"—is similar to the caveats expressed by other leading antirace thinkers, including Du Bois in *The Health and Physique of the Negro American*. That such a qualification was part of the attack on racial science suggests the tentative nature of this kind of thinking (after all, antirace thinking in a scientific guise was really barely a decade old at this point), and also the tentative nature of scientific thinking, which is almost always qualified.[12]

What makes *The Health and Physique of the Negro American* unique is that at a moment when the concept of race was being appropriated by science (in the service of racism), Du Bois was the first to synthesize a growing anthropologic literature arguing that race was not, in fact, a useful scientific category and that race was instead a socially constructed concept. Du Bois accomplished this by logically and rhetorically attacking the idea of race and by backing up these statements with scientific evidence of his own. Du Bois built his antirace concept argument in the following way. First, he began the book by writing that Americans talk about race in the way that they do because they know no better, and he centered his argument in the realm of scientific thought. Americans, Du Bois argued, "are eagerly and often bitterly discussing race problems" because they are behind the scientific times; they are not up-to-date on scientific advances regarding the understanding of human diversity.[13] Second, Du Bois directly attacked the race concept, but he did so by first attacking the idea of whiteness, quoting the anthropologist William Ripley, who wrote that "it may smack of heresy to assert, in face of the teaching of all our textbooks on geography and history, that there is no single European or white race of men; and yet that is the plain truth.[14] Du Bois also cited contemporary anthropological discourse to argue that Europeans were an intermediate people between the Asiatic and Negro races. While some anthropologists would have accepted such assertions, these ideas would have been offensive to many scientists at the time and odious to the vast majority of Americans.

Third, Du Bois extended his attack on race to the idea of a discrete black race generally and of an African American race specifically, writing, "The human species so shade and mingle with each other that . . . it is impossible to draw a color line between black and other races." To this Du Bois adds, "But in all physical characteristics the Negro race cannot be set off by itself as absolutely different."[15] Du Bois also calls into question the idea of a "pure Negro," highlighting that "the Negro-American represents a very wide and thorough blending of nearly all African people from north to south; and more than that . . . a blending of European and African blood."[16] Du Bois acknowledged that neither of these facts would be readily forthcoming in the United States. That because America was a racist society, "no serious attempt has ever been made to study the physical appearance and peculiarities of the transplanted African or their millions of descendants," and that a racist society limits the way in which questions of race themselves could be studied.[17] Or, as Du Bois himself put it, "scientific research seldom flourishes in the midst of a social struggle and heated discussion."[18] By arguing against common assumptions about how humanity was divided, Du Bois sought to bring attention to how human difference existed on a spectrum across the globe and could be organized in different ways.

Fourth, and finally, Du Bois attacked the race concept by examining quantitative data about (alleged) racial differences. Drawing on census data, public health data, military recruitment data (including data from records for U.S. Army recruitment, from the U.S. Census, and from the Freedmen's Inquiry Commission of 1863), and data and conclusions from previously published studies (some of which were used in *support* of the race concept), Du Bois cobbled together a vision of race and of African Americans contrary to how most Americans were thinking. In answer to the claims by nineteenth-century anthropologists, polygenists, and craniologists that Africans and African Americans had a smaller cranial capacity and hence inferior intelligence, Du Bois presented evidence that no differences in brain size or structure had been proven, and that variability within races is similar.[19] Also, in looking at other physiognomic data, Du Bois acknowledged the significant variability within and between races.[20] Finally, based upon analysis of his data, Du Bois concluded that disparities were *not* rooted in biological differences between groups. "If the population were divided as to social and economic condition, the matter of race" in predicting health disparities, Du Bois argued, "would be almost entirely eliminated." Ultimately, poverty and "the conditions of life," as Du Bois called them, were the real causes of health disparities. Du Bois believed that with "improved sanitary condition, improved education and better economic opportunities, the mortality of the race may and probably will steadily decrease until it becomes normal." Drawing on his own research on Philadelphia's African American community, Du Bois argued that differences in health between groups were "an index of social conditions" and anticipated what in today's public health parlance would be called the social determinants of health—including poor nutrition, lack of access to clean water, and low socioeconomic status—being at the root of the disparities he identified.[21]

Several of Du Bois's arguments against the race concept were made in direct response to the assertions of Frederick Hoffman, a statistician for Prudential Life Insurance Company and author of the racist tract *Race Traits and Tendencies of the American Negro*, published in 1896. Hoffman's work both embodied the fading ideas of social Darwinists ("the downward tendencies of the colored race" were leading it toward "gradual extinction"[22]) and the ascendant ideas of eugenics ("given the same conditions of life for two races, the one of Aryan descent will prove the superior, solely on account of its ancient inheritance of virtue and transmitted qualities"[23]). Though not a biologist, Hoffman marshaled statistics to "prove" a biological point in the service of the insurance industry—that black Americans were biologically inferior, that their racial traits consigned them to poverty and unhealthy living conditions, and that they therefore were uninsurable.[24] He set out to make this point following passage of insurance antidiscriminatory statutes in several states. A Massachusetts law was the first, passed in 1884. Others, passed in the following decade, forbade insurance companies from discriminating in their distribution of benefits—blacks could no longer be given fewer benefits than whites if paying the same premiums.[25]

Du Bois had earlier answered Hoffman's racist assertions in an 1897 review of the book in the *Annals of the American Academy of Political and Social Science*. There he attacked *Race Traits* and its conclusions as having "doubtful value, on account of the character of the material, the extent of the field, and the unscientific use of the statistical method."[26] In *The Health and Physique of the Negro American* Du Bois responded more directly. The title—*The Health and Physique of the Negro American*—is an unambiguous retort to the title of Hoffman's book. The two descriptive adjectives in Hoffman's title, "traits" and "tendencies," are suggestive of something negative, whereas "health" and "physique" in Du Bois's title sound positive and healthful. Also, "traits and tendencies" indicates something inborn and permanent, while "health and physique" are things that are experienced and shaped by human beings, or "the conditions of life," as Du Bois called them.[27] By providing data illustrating the relationship between poverty and morbidity and mortality rates, Du Bois rejected Hoffman's claim that blacks are "inherently inferior in physique to whites," and that diseases such as tuberculosis, identified by Hoffman as a black disease, "is not a racial disease but a social disease."[28]

With the advent of eugenics, racial science, which had for most of the nineteenth century been driven by work in anthropology, quickly became largely the domain of biology and genetics. During the course of the twentieth century many anthropologists, led by Boas, would move away from a biologized race concept just as biology and genetics embraced it.[29] While Du Bois could not have predicted this (after all, the term "genetics" was coined by the biologist William Bateson the same year that *Health and Physique of the Negro American* was published), his writings suggest that he sensed what was to come for science and race in the twentieth century, and through this work he had hoped that intellect would prevail over ignorance and that the biologizing of race would be stopped. Looking back on this period in his life several decades later, Du Bois recalled that he "had too often seen science made the slave of caste and race hate."[30] Yet even as Du Bois set out to fight racism through scientific study, rationality, and a belief in justice, events in his own life and in the world around him would force him to rethink that course. In late September 1906, just a few months after the conference that launched *The Health and Physique of the Negro American*, Atlanta became ground zero for a brutal race riot, sparked by local newspaper reports of alleged assaults of white women by black men and caused by underlying racial tensions driven by the consolidation of and resistance to Jim Crow in Atlanta. The white mob was enormous—approximately 10,000 men, women, and children took to the streets.[31] By the time the riot was quelled a few days later, over twenty black residents of Atlanta had been murdered, and many scores more had been beaten.[32]

After the riots, according to historians Dominic Capeci Jr. and Jack Knight, Du Bois "struggled with the profound impact of white brutality on his psyche and philosophy" and "his reliance on rationality."[33] Though Du Bois spent his lifetime fighting oppression, in the wake of the Atlanta riots his approach and style shifted. He would quickly abandon his faith in education, rationality, and science as an antidote to white violence and racism. The approach of Du Bois the social scientist or "organizational leader," as Capeci and Knight call him, had not worked and gave way to Du Bois as "race propagandist" and "neoabolitionist," who came to rely on rhetorical flourish and political effort in his lifelong fight against racism.[34] In this role Du Bois emerged as the leading political voice for African Americans. Du Bois's biographer David Levering Lewis explains that "his chosen weapons were grand ideas propelled by uncompromising language" about America's failings and successes.[35] Du Bois also grew increasingly skeptical of science as an arbiter of truth and the power of truth to change hearts and minds.

In the wake of this personal transformation Du Bois came to believe that racism was "not based on science, else it would be held as a postulate of the most tentative kind, ready at any time to be withdrawn in the face of facts." Racism was instead simply a "passionate, deep-seated heritage, and as such can be moved by neither argument nor fact. Only faith in humanity will lead the world to rise above its present color prejudice."[36] And so, despite his rational, science-based refutation of the biological race concept and of racial science more broadly, an increasingly cynical (or perhaps practical) Du Bois recognized that eugenics would triumph for the foreseeable future, building a biological and hereditarian argument for racial difference that flourished in scientific and popular thought in America. Yet when scientists did begin to rebuke race, they did it largely on the terms that Du Bois had laid out in his elegant study in 1906.

To be sure, throughout his career Du Bois would keep fighting against white supremacy in the guise of racial science and the social damage and intellectual anguish it wrought. Beginning in 1910, over the course of his almost twenty-five-year tenure at the helm of *The Crisis,* the NAACP's official publication, Du Bois used his editorial authority to assail racial science. A 1915 "Opinion" restates a prominent theme from *The Health and Physique of the Negro American,* calling attention to the idea that "no race, as we know races, is an unmixed race. All so-called races are the result of mixtures."[37] A 1911 editorial stated "America is fifty years behind the scientific world in its racial philosophy." The editorial implied that biology and race were divergent; "the deepest cause of misunderstandings between peoples is perhaps the tacit assumption that the present characteristics of a people are the expression of permanent qualities."[38]

Du Bois would also twice debate the eugenicist Lothrop Stoddard. In 1927 Du Bois and Stoddard squared off on radio from New York City. They met again in 1929, this time in a public debate in Chicago with more than 4,000 in attendance. During the raucous debate Du Bois attacked the core beliefs and contradictions of white supremacy. Where "Nordics" like Stoddard sought to maintain racial purity, Du Bois noted that so-called Nordics had, through "exploitation," "spread their bastards to every corner of land and sea."[39] When Stoddard maintained that segregation based on "racial difference" was "not discrimination; it is separation," Du Bois attacked the very notion of race itself, harking back to his work in

*The Health and Physique of the Negro American.* Du Bois sarcastically noted (given his own mixed-race ancestry) that he was "'gladly . . . the representative of the Negro race,' but was also equally capable of being 'a representative of the Nordic race.'" Explaining the contradictions of America's racial mores, Du Bois noted that when white supremacy denied his humanity it did so because of the color of his skin, but when he was deemed worthy in any way, most especially in areas of his intelligence, he was reminded of his white ancestry.[40]

By the 1930s Du Bois was not alone in his attack on racial science and on the race concept itself. He was joined that decade by a growing chorus of natural and social scientists who would, to varying degrees, embrace his stance on the matter. Yet while most of Du Bois's writings expressed an unambiguous rejection of the biological race concept, there are contradictions on the topic in his writing. Despite his rebuke of eugenics, Du Bois, to some degree, embraced the spirit of that movement in his own work. Some have even suggested that he embraced a progressive eugenics, most evidenced in his writings on racial uplift. In fact, that W. E. B. Du Bois continued to struggle with shedding vestiges of the race concept points to how deeply embedded race was in America. That Du Bois implicitly understood the dangers of American racism and, to a great degree, was influenced by one of its most disgraceful racial theories only reinforces this notion.

Nowhere was his eugenic-mindedness more obvious than in his ideal of the "Talented Tenth," a eugenic-sounding "best of the race" who were endowed with the characteristics to lead African Americans. This "metaphorical eugenics," or "bioelitism," as the contemporary critic Daylanne English called it, imbued Du Bois's Talented Tenth "with an explicitly biological superiority."[41] But with an expansive definition of eugenics and heredity, and by cherry-picking examples of an alleged embrace of eugenics from *The Crisis* and selected other writings, it is easy to overstate Du Bois's affinity for such ideas. Ultimately, those who would accuse Du Bois of harboring deep eugenic tendencies confuse his commitment to a class-based and elitist solution to the challenges facing African Americans with a bioelitist one.

Du Bois's antieugenic thinking on the race concept was evident throughout his career. For example, in his 1938 book *Black Folk: Then and Now*, he wrote, "No scientific definition of race is possible," and that "the most that could be asserted of race was that 'so far as these differences are measurable they fade into one another.'" But at the same time, Du Bois biographer David Levering Lewis observes that in that same book Du Bois also engaged in racial essentialism, writing that Negro blood is "the basis of the blood of all men." This contradiction was, according to Lewis, intended to "function in the service of racial pluralism, for in validating an unknown and remarkable Negro past he envisaged a future in which all races could accept the cultural parity of one another's history as well as the interdependence of their destinies." Lewis concludes that Du Bois's "rather mild racial essentialism" was itself an attack on contemporaneous race supremacy dogmas.[42] Still others have noted that even as Du Bois's rejection of a scientific race concept became more strident (in a *Crisis* article in 1911 and in his 1940 autobiography), this explicit rejection remained complicated by Du Bois's pursuit of a unifying concept for persons of African descent in a racialized world.[43]

## HISTORICIZING RACE

By the 1930s, as eugenical theory was becoming less palatable to both scientists and the general public, prominent academics joined Du Bois in attacking the race concept. The Columbia University historian Jacques Barzun, author of one of the first in a spate of academic books about race and racism published during the 1930s and early 1940s, called race "one of the great catchwords about which ink and blood are everywhere spilled in reckless quantities."[44] Published in 1940, *Race: Science and Politics* by the anthropologist Ruth Benedict attacked racism, but noted "that to recognize Race does not mean to recognize Racism. Race is a matter of careful scientific study; Racism is an unproved assumption of the biological and perpetual superiority of one human group over another."[45] Benedict called race "a classification based on traits which are hereditary."[46] She belonged to a group of scholars, including the sociologist Robert Park and the economist Gunnar Myrdal, who, rather than emphasizing problems with the race concept, looked instead to race relations as playing the functional role in America's racial calculus.[47] What made Benedict such an important part of this movement were her assertions that science was not to blame for the problems of racism; politics was. She wrote, for example, that in order to understand race persecution, we do not need to investigate race; we need to investigate *persecution*.[48] While there is a certain logical truth to this position, it abdicates scientists of the responsibility of how the race concept was utilized popularly and it ignores a long history of scientists supporting racism through their work. Benedict acknowledged as much, writing that "for the scientist, science is a body of knowledge; he resents its use as a body of magic."[49]

Barzun's 1936 text *Race: A Study in Modern Superstition*, like Du Bois's work thirty years earlier, utilized both historical methods and the latest science to rebuke the race concept. A budding young historian at Columbia University, Barzun tackled the subject of race as American eugenics was in decline, the smoldering racial hatred of Nazism was fast making an impact on the European continent, and the movement for civil rights for African Americans was is in its earliest days.[50] For Barzun the only possible argument in favor of the race concept "would be that no one race could possibly have been gifted with such a capacity for nonsense as the literature on the subject affords."[51]

While Barzun's interest was primarily in considering "racialism as a European phenomenon"—the book is infused with anticipation and fear of what Nazi racism was wreaking in Europe, from where Barzun had hailed—it also conveys an implicit understanding of how racial thought impacted African Americans.[52] Barzun writes that "those human beings who have not lost their pigmentation are simply more clearly marked than others for discrimination; they wear a uniform that they cannot take off." This social stigma, which Barzun points out was not unique to the American experience, highlights the fact that "the problems of colored populations . . . are not problems about a natural fact called race: they are problems of social life, of economic status, of educational policy, and of political organization."[53] Barzun also believed that the "race thinking," as he called it in a letter to the sociologist Oliver Cromwell Cox, was endemic to Americans' thinking across social, cultural, and economic spectrums. In Barzun's estimation race thinking applied "to taxi drivers as well as to Nordics or Semites. In fact it will fit any labeled category. It is as common among football fans as among rabid nationalists and as such deserves a place as the root fallacy in all social antagonisms that are not based on direct competition for concrete goods."[54]

Barzun, certainly one of the first authors to historicize the race concept, would go on to have a long and distinguished career as a historian. *Race*, his first book, was in general not reviewed kindly in either the social or natural scientific literature. Several reviewers acknowledged Barzun's contribution in providing a broader context for the history of race but questioned his rejection of the race concept. Writing in the journal *Man*, the British social anthropologist Rosemary Firth wondered if "to deny any scientific reality to race at all . . . is to carry the valuable indictment of the popular absurdities of race thinking to an extreme, and thereby somewhat to weaken the otherwise good case" against its misuse.[55] Even the *American Historical Review*, a distinguished journal for American historians, allowed the anthropologist Clark Wissler of the American Museum of Natural History to review the book. Wissler, a member of the eugenicist Galton Society, was a steadfast believer in the biological race concept and an advocate of Nordic superiority. He argued that the main sources of human evolution were from Asia and Europe, "while the rest of the world was relegated to a marginal position."[56] Wissler was not particularly kind, writing, "Though the author [Barzun] regards this volume as a critical history of thought on race, it can hardly qualify as a calm weighing of evidence."[57] Wissler dismissed Barzun's work as extreme and suggested that it was best read in conjunction with books "which defend the concepts of race purity." Such an exercise, Wissler suggested, "may give perspective" on the subject. Barzun, angered by Wissler's review, sent a letter to the editor of the *American Historical Review* defending the book and pointing out Wissler's errors and misconceptions. Barzun seemed particularly irritated at the way Wissler, writing in the leading historical journal no less, criticized *Race* and defended the race concept by demeaning the historical craft and the ability of historical methods to evaluate science. In the letter to the editor, Barzun pointed out that "the basic questions he raises without stating them seem to be: whether ideas are real forces or mere illusions; whether their relation to biological or economic fact are subtle and complex or obvious and simple (Race being a simple First Cause for culture) and ultimately, whether the student of ideas has the right to be a pragmatic critic instead of a mechanistic materialist."[58]

*Race* concluded with twelve objections to the race concept, which, very much in the mold of Du Bois's earlier critique, included the concept's general inconsistency, its elusiveness, its statistical fallacy, its fallacy of genetic predetermination, and its absolutism. Race could explain anything and everything, and Barzun noted that "in a real world of shifting appearance, race satisfies man's demand for certainty by providing a small, simple, and complete cause for a great variety of large and complex events."[59] Barzun's book and its critique of the race concept remain historiographically important for telling a modern reader where to look for the changing nature of the scientific foundation of the race concept. If anthropology was the arbiter of meaning for race in the nineteenth and through the earliest years of the twentieth century, then genetics, or "the problem of hereditary transmission," as he called it, is "central in any future theory of race."[60]

Barzun could claim genetics as the emerging authority on the race concept with confidence for several reasons. First, political events in Europe, specifically the rise of Nazism, had helped popularize the link between race and genetics in a way that not even the most fervent American eugenicists had thought possible—although they actively and without compunction sought out this role. To a significant degree, Nazi eugenic zeal was inspired by American eugenics. The publication of Madison Grant's eugenic tract *The Passing of the Great Race: The Racial Basis of European History* might have preceded the rise of Nazism by more than a decade, but its ideas about Nordic racial purity influenced many Germans.[61] In a letter to Grant, Hitler called *The Passing* "his Bible."[62] In 1933 the *Eugenical News*, the official newsletter of several eugenic organizations including the American Eugenics Society, noted the American influence on German sterilization policy: "To one versed in the history of eugenic sterilization in America, the text of the German statute reads almost like the American model sterilization law."[63] American philanthropists, including those of the Rockefeller Foundation, also gave scientific grants to German eugenicist researchers, both before and for several years after the rise of Hitler. And even as the world recoiled in horror at the ways in which the Nazis integrated eugenics into their political philosophy— mass sterilizations and concentration camps—American eugenicists continued to support their Nazi brethren. In 1935 Harry Laughlin accepted an honorary degree from the University of Heidelberg for "being one of the most important pioneers in the field of racial hygiene." The dean of the University of Heidelberg's medical school later helped organize the gassing of thousands of mentally handicapped adults. Also in 1935, after a visit to Berlin, the head of the Eugenic Research Association, Clarence Campbell, proclaimed that Nazi eugenic policy "sets a pattern which other nations and other racial groups must follow if they do not wish to fall behind in their racial quality, in their racial accomplishments, and in their prospects for survival." Finally, in 1937, American eugenicists distributed a Nazi eugenic propaganda film to promote the eugenic cause in the United States.[64]

Second, with the American eugenics movement in decline during the 1930s, there was an opportunity for new claims and room for new authorities on the biological race concept. Third (and this is more speculation than assertion), as a historian, Barzun was able to see the evolution of racial science—from its origins in prescientific folkways to anthropological to biological to genetic thinking—having covered more than a century of its presence in European and American social and intellectual thought in his book *Race*. Following the debates, discussions, and findings on race, Barzun would have seen that geneticists were quickly becoming the respected voice on this matter. Fourth, and finally, the changing technological, methodological, and ontological approaches in biology, evolutionary biology, and genetics were coalescing into what in the 1930s would become known as the evolutionary or modern synthesis and reshaping the way the biological sciences hypothesized, conceptualized, and analyzed human difference. It was from this modern synthesis that ideas about race would be reformulated into both the scientific and popular lexicons.

# 7

# RACE AND THE
# EVOLUTIONARY SYNTHESIS

The evolutionary synthesis in biology, a historic moment for the biological sciences, was the union, in a Darwinian context, of theoretical population genetics, experimental genetics, and natural history.[1] The synthesis resolved several outstanding, and until that point seemingly intractable, issues for biology. First, it reached accord that natural selection was the mechanism that accounted for evolutionary change. This also meant the acceptance of evolution as a gradual process. Second, the synthesis resolved the long-standing issue of how to explain evolutionary phenomena. According to Ernst Mayr, himself an architect of the synthesis, the consensus that emerged on this issue understood that "by introducing the population concept, by considering species as reproductively isolated aggregates of populations, and by analyzing the effect of ecological factors . . . on diversity and on the origin of higher taxa, one can explain all evolutionary phenomena in a manner that is consistent both with known genetic mechanisms and with the observational evidence of the naturalists."[2] Consensus on these issues was, to be sure, difficult for all involved. For example, for the first third of the twentieth century systematists rejected the discoveries of genetics and held fast to a worldview shaped by their own discoveries. This was primarily because their concepts of variation and inheritance were consistent with their own field observations and, they believed, could explain evolution better than the Mendelians.[3] But between 1936 and 1947 this synthesis finally emerged in the literature, bridging a gap between two areas of research that had up until this time failed to communicate with each other.[4]

Theodosius Dobzhansky, one of the architects of the synthesis, described the significance of the emerging consensus in a 1947 letter to R. C. Murphy, president of the American Museum of Natural History: "It is not particularly inspiring to continue to taxonomize as the nineteenth century taxonomists did. But no informed person can disregard the fact that a new science is being born from a synthesis of morphological and experimental biology, and that this science begins to occupy one of the central positions among the biological disciplines. On the taxonomic side, the new systematics uses, in part, the same apparatus of research which the old taxonomy has assembled and used, namely museum collections of dead specimens. But the new systematics studies dead specimens not merely in order to arrange them on museum shelves; dead specimens are used as means to discover the new laws of life."[5] Among the seminal works of that period were Dobzhansky's *Genetics and the Origin of Species*, Julian Huxley's *Evolution: The Modern Synthesis*, George G. Simpson's *Tempo and Mode in Evolution*, and Ernst Mayr's *Systematics and the Origin of Species*.[6]

Ernst Mayr, one of the architects of the evolutionary synthesis, has written about the impact the synthesis had on a biological understanding of race and human diversity. According to Mayr, one of the more significant characteristics of presynthesis thinking was that while the "naturalists had wrong ideas on the nature of inheritance and variation; the experimental geneticists were dominated by typological thinking that resulted in pure lines and mutation pressure."[7] Mayr believed that prior to the synthesis geneticists were stuck in a mode of thought in which "species and populations were not seen as highly variable aggregates consisting of genetically unique individuals, but rather as uniform types."[8] This perhaps paints too simplistic a picture of broader arguments about the nature of variation in presynthesis biology. Mayr's point makes the most sense, however, not in explaining the fixity of two contrasting worldviews (typology versus populationist thinking) but as a way to illustrate the way typology had, before the synthesis, circumscribed thinking about variation.

And Mayr understood these contrasting worldviews to be a fundamental struggle within evolutionary thinking; in fact, his writings traced the origins of modern populationist thinking back to Darwin and his cousin Galton. Galton, while the epitome of a typologist, would contribute to the emergence of populationist thought by developing mathematical models that accounted for variability.[9] The sharp dichotomy between pre- and postsynthesis thinking on variation as Mayr described it was not, of course, absolute, and others have acknowledged the transitional work on variation occurring in genetics research in the decades just prior to the synthesis. For example, in the work on human ABO blood-group maps from the latter years of the first decade of the twentieth century through the 1940s, one can see the evolution from typological to populationist thinking.[10]

But regardless of this transitional work in human diversity, the failure to fully conceptualize genetic variation led directly to the belief in the existence of "uniform types," or races as they were commonly known. One of the more significant effects of the synthesis would be the rejection by most biologists of typological thinking; it may be recalled from chapter 5 that one of the reasons why eugenics fell out of favor, and why Charles Davenport's status in American science quickly went from leader to quack, was his failure to embrace this shift.[11]

Yet this rejection of typology by the "new" biology did not mean that an evolving biological race concept somehow left scientists in the field above racist conjecture or left the race concept itself invulnerable to misuse. Indeed, there is no doubt that while by the 1930s new findings in population genetics and evolutionary biology witnessed significant changes in racial thinking, there was still no shortage of opportunity for the integration of racism into biological conceptions of human difference.[12] To be sure, there remained those who held on to the legacies of typological thinking who were part of the modern synthesis. Julian Huxley was, for example, *the* popularizer of the synthesis and the author of a book on race titled *We Europeans*—written in large part to attack the Nazi's racial doctrines, the book rejected the race concept as unscientific. He would also write in 1938 that "in human genetics, the most important immediate problem is to my mind that of 'race crossing.' . . . The question whether certain race crosses produce 'disharmonious' results needs more adequate exploration. Social implications must also be borne in mind in considering this subject."[13]

Despite Huxley's prominence in the synthesis, his typological "voice" on this subject becomes less interesting than that of persons who through their research, publications, and correspondence struggled to redefine the postsynthesis biological meaning of race difference in humans. Ultimately, it would not be the holdover typologists and racists who developed a modern biological race concept. It was rather the scientists who sought to reconcile concepts of human diversity and modern biology who shaped a new consensus on racial difference. Those who sought to eliminate the race concept in biology altogether also played an important role in this debate. It is more interesting to consider those who recognized the need for change in the context of the new synthesis and how they made that change rather than to examine "more of the same" racist conjecture. It is the legacy of this work on race and human genetic diversity from the 1930s through the 1960s that has had continued significance as modern biology, genetics, and related disciplines continue to struggle with these concepts. And it is the work in this area conducted by the evolutionary biologist Theodosius Dobzhansky, the population geneticist L. C. Dunn, the evolutionary anthropologist Ashley Montagu, and the geneticist Curt Stern that will capture the attention of the remainder of the present and subsequent chapters.

## THEODOSIUS DOBZHANSKY AND THE CREATION OF A NEW RACE CONCEPT

Theodosius Dobzhansky, according to the evolutionary biologist and geologist Stephen Jay Gould, "was preadapted to initiate the synthesis."[14] A Russian-trained zoologist who received a grant from the Rockefeller Foundation to study with Thomas Hunt Morgan beginning in 1928 (first at Columbia and then at Caltech), Dobzhansky's training in Russia and subsequent work in the United States prepared him for his role in the modern synthesis in a way that was unparalleled in biological circles at the time.[15] From Russia Dobzhansky inherited a rich tradition of experimental genetics in the context of natural history, a tradition absent at that time in American biology.[16] Dobzhansky would later recall that to Morgan, "'naturalist' was a word almost of contempt . . . the antonym of 'scientist.'"[17] Nevertheless, in the United States Dobzhansky had the opportunity to apply this tradition by working with Morgan and others. Surrounded by such a distinguished group of colleagues, Dobzhansky thrived and quickly went on to become one of the architects of the evolutionary synthesis and one of the foremost evolutionary geneticists of the twentieth century.

The importance of Dobzhansky's work and those biologists who helped drive the evolutionary synthesis forward is well documented. One historian, for example, underscores the significance of Dobzhansky's *Genetics and the Origin of Species*, writing that it "thus became the ground on which the heterogeneous practices of the biological sciences were stabilized and bound."[18] At the time of its publication the population geneticist Sewell Wright, whose own mathematical models proved indispensable to Dobzhansky, heralded *Genetics and the Origin of Species* as "a book which will be a necessity for all interested in the recent development of the theory of evolution" and called it "by far the best synthesis that has come out."[19] In an introduction to the 1982 edition of the book, Stephen Jay Gould calls it "a long argument for a general attitude toward nature and a specific approach that might unite the disparate elements of evolutionary theory."[20]

The historiography on racial science—the murky place where the biological race concept intersects with prejudice both within and outside science—suggests that in post–World War II America the race concept continued a decline it had begun in the 1920s. According to Elazar Barkan, that decline ushered in an era that saw the emergence of egalitarianism in scientific discourse that would force racial science into retreat.[21] This book argues instead that racial science—if defined as the use of science, both by scientists and laypersons alike, generally as part of a greater arsenal of oppression, and specifically as a scientific language utilized for this general purpose—did not decline. Instead, in the interwar period racial science thrived, and beginning in the 1930s through the postwar period, the language and ideas of racial science were adapted to the evolutionary synthesis by both participants in the synthesis and by outsiders trying to influence debates on the nature of human genetic differences. This is not to say that the evolutionary synthesis was itself racist in the same way as the eugenics movement. Indeed, most modern geneticists rejected racism in the 1940s but nevertheless continued to embrace the biological race concept.[22] This seeming contradiction was at the center of efforts by many geneticists to preserve the race concept for their work (in all species) and cast off the racist aspersions formerly associated with it. As the nation's preeminent geneticist, and as one of the architects of the evolutionary synthesis in modern biology, Theodosius Dobzhansky was at the forefront of this change.

From Dobzhansky's perch as one of the preeminent scientists of his time he had the desire, prestige, and audience to write on matters social as well as scientific. And so he did, reaching beyond the confines of academia on subjects ranging from race and society to radiation and its impact on humankind to popular books about human evolution.[23] For an émigré from Russia who had begun learning English only upon his arrival in the United States, this was quite a feat. His first popular book, *Heredity, Race, and Society*, published in 1946, was written on the subject of race and coauthored with his colleague at Columbia (where he had become a professor of zoology in 1940), the geneticist L. C. Dunn. But Dobzhansky had already written about the race concept in biology—with a very specific purpose—earlier in the 1940s, and his scientific work on the modern synthesis as well as his deep humanitarian streak motivated his work in this area.

The laboratory life of Theodosius Dobzhansky had a marked impact on the way he perceived and conceptualized difference in the species he studied and, by extension, on how he understood diversity in his fellow man. The fact that Dobzhansky's scientific and intellectual authority as a leader of the evolutionary synthesis was brought to bear on the development of a new race concept in biology should not be surprising given the importance of genetic variation as a key concept in the synthesis. For Dobzhansky and others involved in the synthesis, natural selection, speciation, and adaptation could be understood only in the context of population, not typological thinking, and understanding genetic variation within populations was essential to understanding the mechanisms of evolution. Dobzhansky's Russian training played a formative role in his approach. In Russia, where Dobzhansky began his training as an entomologist, his work was rooted in a system of knowledge that emphasized natural history in a way that the American and European biological traditions did not.[24] And it is in Dobzhansky's earlier foundational work on variation, race, and speciation that these Russian roots are evident. While still in Russia, prior to his work with *Drosophila pseudoobscura*,

Dobzhansky worked primarily with ladybug beetles (*Coccinellidae*) as his experimental animal of choice. One early paper on ladybugs sought to examine the problem of "the formulation of races by means of selection of biotypes." Other papers would look more closely at the mechanisms and nature of speciation, and during the 1920s Dobzhansky's research program matured to examine "geographical distribution and variability in populations, individual variability within populations, and inheritance of variability"—all important components of his growing research into microevolution. Although the seeds of Dobzhansky's synthetic thinking would be present in his 1920s work, at that time he still had yet to make the connections between evolution, genetics, and morphology that would be present in his groundbreaking work of the 1930s.[25]

It is not simply that Dobzhansky was interested in problems of race and variation in his earlier work; his training as a naturalist helped to focus his understanding of the genetics of natural populations. The attention to populations in the wild, as opposed to laboratory-bred populations, was a characteristic of the Russian biological tradition emphasizing the naturalist's viewpoint—one that was centered on the observations of wild populations and using those observations to construct wider theories about the nature of biological systems at the micro- and macrolevels.[26] Dobzhansky ascribed the roots of this tradition in Russia to "perhaps the great size and environmental diversity of the country," which led to "new and unusual animals and plants [that] were collected and brought for study in university laboratories, zoological and botanical museums and marine institutes. A majority of biologists had experience working in the field and observing living beings in their habitats. . . . Anyway, it was taken for granted that a biologist must know animals, or plants, or both."[27]

By fusing experimental and laboratory traditions in evolutionary biology, Dobzhansky was able to not only speculate about the nature of populations through laboratory analysis but also, through his meticulous fieldwork in natural environments, witness the genetics of population dynamics firsthand. Through his long-standing collaboration with the population geneticist Sewell Wright, Dobzhansky was able to make sense of the genetics of natural populations. Dobzhansky provided the observational and biological data as well as the theory; Wright helped Dobzhansky with the mathematical models underlying those theories.

The breaking down of barriers between experimental work (the laboratory geneticist) and fieldwork (the naturalist) was, of course, another important legacy that Dobzhansky left to biology.

Dobzhansky's description of a 1943 trip to Brazil in a letter to Sewell Wright suggests the way his fieldwork shaped his own understanding of variation in natural populations. "Tropical forest is an environment so different from ours of the temperate zones that all biological conceptions or preconceptions with which a 'temperate' biologist starts have to be critically examined," Dobzhansky wrote. Dobzhansky noted that "the richness of species is of course commonplace; the population density of most species being or seeming low is also something for what I have been prepared." But he was surprised to "see the incredible heterogeneity within even what seems superficially the same type of environment. Up to now I can not tell with assurance which *Drosophila* species is the commonest because different ones are commonest in different places."[28]

Race as a taxonomic term predated Dobzhansky's reimagining of the concept in biology and would have been a classificatory tool used by taxonomists for at least a century before the evolutionary synthesis. Race was part of the set of standard categories that Linnaean taxonomy utilized to help organize biodiversity, specifically the morphological and physiological diversity found within named species. As a Russian trained natural scientist, Dobzhansky utilized race in his work to describe species subdivisions, or subspecies. In fact, race and subspecies were often used interchangeably in the naturalist literature. Dobzhansky described the taxonomic meaning of race in a 1933 paper on "Geographical Variation in Lady-Beetles." "The different patterns known in a given species do not occur equally frequently in every part of the area inhabited by the species," Dobzhansky wrote about the variation of color patterns found in lady beetles. Dobzhansky continued, "In some sections of the specific area a majority of the population may consist of individuals having a pattern or patterns which are rare or absent in other sections of the same area. The species becomes, thus, differentiated into geographical races (subspecies). Each of the subspecies is characterized by a definite frequency of the different patterns in the population."[29]

It is no accident that Dobzhansky, who was trained in this tradition of natural history, was able to make the theoretical leaps that helped usher in the evolutionary synthesis. It is also no accident that, like the Russian colleagues he left behind, he was mathematically unsophisticated, a shortcoming he himself acknowledged. Writing to his collaborator and mathematical muse, the theoretical population geneticist Sewell Wright, Dobzhansky admitted his own "mathematical understanding is far too insufficient to read and understand" Wright's papers "completely." "But," Dobzhansky continued, "I have done the same thing that I have with other papers: read the part of the text preceding and following the mathematics, skipped the latter in assurance that to it the expression 'papa knows how' is applicable." Throughout his career Dobzhansky turned to mathematically minded population geneticists for guidance on the bases of his evolutionary theories. Fruitful collaborations with Alfred Sturtevant and Sewell Wright were critical to his overall success as a scientist.[30]

But Dobzhansky, of course, was not the only evolutionary synthesizer to write about variation in the scientific literature. Sewell Wright and Ernst Mayr, for example, wrote extensively about genetic variation in the 1930s and 1940s but paid only minimal attention to the race concept.[31] Ultimately, that Dobzhansky became so invested in discussions about genetic variation in the context of the synthesis and in both scientific and popular deliberations about the race concept had also to do with his personal history as well as his political and moral beliefs. Dobzhansky's thinking on race was shaped by both his deep-seated humanism and his belief in the important role of the intellectual in society. These forces drove his scientific interests. Furthermore, Dobzhansky's dedication to the study of evolution was not simply about uncovering the mechanisms of that process but was also about understanding the implications of evolution in and for humans—an approach that shaped his popular discussions of race.[32]

Looking back on his career and on his role in scientific debates about race, Dobzhansky's own hindsight suggests that a veritable "changing of the guard" in his field (older geneticists retiring and dying) precipitated his and others' entry into the debates on race. Dobzhansky recalled that race prejudice was potent among the "older geneticists" and remembered an exchange with Edward East while visiting Woods Hole Laboratory in Massachusetts in 1936. East insisted that Dobzhansky was "not a Russian" given that Russians are a genetically inferior people. A small minority of Nordics lived among that population, and Dobzhansky had to have been part of that group. Dobzhansky was also motivated to become involved in debates about the nature of the biological race concept because of the "impact of Nazi atrocities against the Jews" and because the biological evidence did not support the alleged dangers of race crossing.[33] Dobzhansky would have seen those who held fast to racist and typological viewpoints as anachronisms, and he himself believed that "changing one's views is not at all shameful, and in fact when we reach the state when we can no longer change it may be time to retire."[34] But it is clear that it was not simply that Dobzhansky's interest in race was about a temporal changing of the guard but also about significant differences in the worldviews of those who chose to tackle the race issue beginning in the 1930s.

The other important characteristic that defined Dobzhansky to a very large degree, and also defined others involved in rethinking the race concept in biology, was that they were outsiders of one sort or another. Dobzhansky was not simply an immigrant to the United States but also an exile who could not return to his Russian homeland in the wake of Stalin's growing purges. Dobzhansky was also, early in his career, a scientific outsider, arriving in the United States with only limited training in genetics and a strong background, unlike most of his colleagues, in natural history. Dobzhansky's collaborators and allies on matters of race had similar stories. L. C. Dunn, who briefly flirted with the eugenics movement, had a son with cerebral palsy whose condition would come to shape his antieugenical thinking. Ashley Montagu, born Israel Ehrenberg in London in 1905, changed his name to avoid the anti-Semitism of the British academic aristocracy. And Curt Stern, a German Jew, was unable to return home after the rise of Hitler.

Dobzhansky's objective—seeking to redefine the biological concept of race in a genetic context—began in earnest in his 1937 *Genetics and the Origin of Species*. While race is not itself the subject of the book, the importance of the concept of variation to the evolutionary synthesis meant that defining populations and other groupings of organisms received significant attention. By privileging variable over fixed populations, Dobzhansky's synthesis helped reconceptualize the idea of difference between and among populations of all organisms, including human beings. His description of the race concept in *Species* should be considered one of the most important writings in the history of this subject, yet it has received relatively little attention. This is surprising given Dobzhansky's standing in genetics, his popular and scientific writings on the subject, and the legacy he left in the form of his students (several of whom, including Richard Lewontin and Francisco Ayala, later contributed to the discussions about race and genetics and were leaders in the fields of genetics and evolutionary biology).

To be sure, Dobzhansky was not the only biologist writing about race at that time, nor was he the first to write against the typological race concept. Indeed, typological approaches to race difference were, over the course of the 1920s and 1930s, losing favor among geneticists as population genetics offered alternative explanations for intraspecies diversity.[35] That information was, albeit slowly, making its way into popular thought about race even as Dobzhansky was helping to devise the evolutionary synthesis and place the race concept within that context. For example, Ralph Bunche, who would later garner fame as a United States representative at the United Nations and as the first African American to win the Nobel Peace Prize (1950), wrote *A World View of Race* the year before *Genetics and Origin of Species* was published. In his book Bunche, then a freshly minted Harvard Ph.D. in political science, wrote a chapter titled "What Is Race?" in which he reviewed the then current state of race thinking in anthropology and biology. In that chapter Bunche described the limits of typological thought ("the plain fact is that the selection of any specific physical trait or set of traits as a basis for identifying racial groups is a purely arbitrary process"), historicized race ("our concept of race is a comparatively recent one"), and reviewed emerging biological ideas about the race concept ("the study of hereditary phenomena, or genetics . . . throws much light upon this human problem of race").[36] In the emerging language of population genetics, Bunche, commenting on why pure human races did not exist, wrote that "human variation is so great, in fact, that perhaps such homogeneous groups never existed at all." Bunche also suggested "we drop the term *race* with reference to existing groups and substitute some more accurate description such as *ethnic groups* or *peoples*."[37] This call to drop "race" went far beyond what geneticists like Dobzhansky were willing to do in rethinking the concept but anticipated what would become a more commonplace suggestion among some anthropologists and biologists in the 1940s and 1950s. Finally, Bunche, like Du Bois before him and Ashley Montagu in his footsteps, concluded that "though racial antagonisms constitute a serious world problem, they have no basis in biology, nor can they be accepted as the inevitable result of group differences. Such differences must be analyzed and understood in their social and historical setting."[38] Unlike Bunche, Dobzhansky would not reject the biological race concept, but did concur that "the term race is one employed . . . with a looseness and inaccuracy matched only by its frequency in our literature."[39] In *Genetics and the Origin of Species*, Dobzhansky hoped to remedy this problem.

A reader first encounters the term "race" in chapter 3 of *Genetics and the Origin of Species*. Dobzhansky wrote that "a living species is seldom a single homogenous population. Far more frequently species are aggregates of races, each race possessing its own complex characteristics. The term 'race' is used quite loosely to designate any subdivision of species which consists of individuals having common hereditary traits."[40] Dobzhansky, however, admitted to complications regarding the use of this term; he was troubled that "so many diverse phenomena have been subsumed under the name 'race' that the term itself has become rather ambiguous." And Dobzhansky worried about the unavoidable contradictions when applying the term, noting if races are "described usually in terms of the statistical averages for all the characters in which they differ from each other," it inevitably "begins to serve as a racial standard with which individuals and groups of individuals can be compared." Dobzhansky recognized the limitations of such a model: "From the point of view of genetics such an attempt to determine to which race a given individual belongs is sometimes an unmitigated fallacy." This is because—and this point is central to the modern synthesis's novel view of race—"racial differences are more commonly due to variations in the relative frequencies of genes in different parts of the species population than to an absolute lack of certain genes in some groups and their complete homozygosis in others. . . . *Individuals carrying or not carrying a certain gene may sometimes be found in many distinct races of a species.*"[41]

By revealing a core contradiction of race in a population genetics context, Dobzhansky explicitly acknowledged the imprecise nature of race. But he didn't stop there. He also noted "that individuals of the same race may differ in more genes than individuals of different races"; that the "units of racial variability are populations and genes, not the complexes of characters which connote in the popular mind a racial distinction"; that to understand race "the geography of the genes, not of the average phenotypes, must be studied"; and that "racial variability of phenotypic traits is continuous."[42] Yet despite Dobzhansky's insights into the limitations of a biological race concept, he refused to jettison it, and in his writings throughout his career, including *Genetics and the Origin of Species*, he sought to articulate why this was so.

In *Genetics* Dobzhansky insisted that the discussions about what he refers to as the "endless and notoriously inconclusive discussion of the 'race problem'" could be boiled down to the following question: "Is a 'race' a concrete entity existing in nature, or is it merely an abstraction with a very limited usefulness?" But for Dobzhansky, this question missed the point: race was neither a concrete or "static" entity nor an abstraction, but instead an organic "process" that can be identified as the frequency of a gene or genes in a segment of a population begins to differ from the rest of that population. Race is therefore a temporal and geographic concept, changing as gene frequencies change, sometimes a stepping-stone to the formation of a new species as genes become fixed in populations as mechanisms prevent interbreeding between races, thereby "splitting what used to be a single collective genotype into two or more separate ones."[43] The caveat to this, of course, is that gene flow must stop between populations for this to happen. If it does not, then, in Dobzhansky's estimate, races will always be in flux and to a large degree indeterminable.

In a paper published in the *Scientific Monthly* in February of 1941, Dobzhansky outlines his justifications for why the biological race concept was as important for studying humans as it was for studying *Drosophila*. The article's content is an expanded view of what he says on the subject in *Genetics*. Dobzhansky argues that previous discussions of the race concept had not been "conducted on a scientific plane," primarily because biologists had "with few exceptions, disdained to take part in the debate." Anthropologists and others had gotten the race concept wrong and he was going to fix it. These claims by Dobzhansky read more like rhetoric than fact, helping him structure an argument that essentially says, *what came before me was scientific bunk proffered by nonbiologists. As an expert biologist I will offer important scientific insight to this debate.* Where

Dobzhansky was correct, however, was in his assertion that biology itself offered "no clear definition of what constitutes a race." In *Genetics*, in the *Scientific Monthly* article, and in many future publications, he sought to offer a working definition that he believed was essential to the work of scientists operating in the paradigm created by the evolutionary synthesis. In Dobzhansky's calculation, "the refined analytical methods of modern genetics may permit a better insight into this problem to be gained than was possible in the past, but the work in this field is now barely begun."[44] And of that past, of those anthropological methods, Dobzhansky recognized their origins—"a strictly pragmatic purpose—to place the almost infinite variety of living beings into pigeonholes where it can be kept until it is needed for further study"—and their shortcomings—"what actually happened was that pigeonholes were sanctified to become God given units (nominalism from which zoological and botanical taxonomies are just now trying to free themselves)."[45]

Dobzhansky began his argument in support of the biological race concept by rejecting the taxonomic and anthropological definition, one that posits races as averages of morphological, physiological, and psychological characters. This typological approach could, in Dobzhansky's estimation, provide a "rough description of the observed variety of humans or of other living beings." It could not, however, provide "an analysis of the underlying causes of this variety." To Dobzhansky, this problem could be fixed by the application of genetics in helping to understand the processes of organismal diversity. As he had previously suggested in *Genetics*, it was not precise enough for geneticists to "define races as populations that differ from each other in the frequencies of certain genes." Ultimately, such a definition tells us little about the "extent" of such differences. Dobzhansky provided an example from his work with *Drosophila* to explain the shortcomings of this definition and provide one of his own. It is worth noting that Dobzhansky anticipated critiques of using *Drosophila* models to discuss human races, writing that "the laws of heredity are the most universally valid ones among the biological regularities yet discovered. The mechanisms of inheritance in man, in the *Drosophila* flies, in plants and even in the unicellulars are fundamentally the same."[46] That Dobzhansky sought consistency in the use of racial terminology across species and clades was driven by his classical training in taxonomy and his rigid view of terminology. Also, as mentioned above, Dobzhansky was trying to appropriate the race concept under the evolutionary synthesis. In order to do so it seems certain that he needed to rid it of its anthropological and eugenic stigma. By placing race under the rubric of genetics, he had hoped to save it for biological practice and thought.

Dobzhansky's *Drosophila* examples were meant to reinforce the biological nature of the race concept. In the *Scientific Monthly* paper Dobzhansky describes how no populations of *Drosophila* are genetically uniform, and that "in every one of them some individuals carried chromosome structures and mutant genes not present in others." He also notes that the genetic composition of populations in this species do not remain constant over time. These changes, as he noted in *Genetics*, can, over evolutionary time, lead to speciation. Changes of this nature are "subject to natural selection." Dobzhansky also notes that none of the populations of *Drosophila* are "pure races." In bringing up this point Dobzhansky was directly engaging the broader social debates about race, and in doing so hoped to disconnect the science from its social context. In Dobzhansky's estimation, "the idea of a pure race is not even a legitimate abstraction: it is a subterfuge used to cloak one's ignorance of the nature of the phenomenon of racial variation." Instead, populations showed variability between them in "geographically graded series" or clines.[47]

One of the challenges Dobzhansky faced using *Drosophila* races as a way to explain the concept for its use among *Homo sapiens* is that there is an obvious distinction in race between human and nonhuman species; the use of race in *Drosophila* obviously does not carry the same social and historical burden as it does for the human species. For naturalists and evolutionary biologists, race was an important piece of the puzzle of natural selection, the chief mechanism of evolution. Species are, of course, made up of heterogeneous populations. That heterogeneity in nonhuman species leads to geographical races (or populations), and these populations are the raw material for speciation. Humans, however, do not have the same ecological niches as other species and therefore are not races in the same way in this evolutionary context.[48] In trying to preserve the race concept for all taxonomic usage, Dobzhansky never explicitly addressed these inconsistencies.

Dobzhansky concluded his article by again arguing that postsynthesis genetics must be the arbiter of the race concept. His rhetoric called for more scientific study into this matter and noted that despite the difficulties involved in race-level research in populations, "the difficulty of the task is not a sufficient reason to cling to the outworn methods of racial study . . . and still less is it a reason for erecting far-reaching theories on the basis of admittedly faulty data. To do so would be a travesty on science."[49] In trying to appropriate race for postevolutionary synthesis genetics, Dobzhansky was trying to preserve a biological race concept. But in his "mission" to inject genetics into the debate over the meaning of race, Dobzhansky was up to something more, and concomitantly hoped that if the public understood the "ABCs of genetics," then the biological race concept would not engender racism.[50] In a 1947 letter to his colleague L. C. Dunn, Dobzhansky referred to several encyclopedia commissions he had recently taken on the subject of race and science. "The pay is not very attractive," Dobzhansky wrote, but "I am told that this book will be in every school library all over the country, so I regard this as a sort of obligation." That obligation, he wrote Dunn, was to "help straighten things out" given "what stuff was fed about human heredity to people." In considering whether to take the commission to revise *Encyclopedia Americana*'s entry on "Races," Dobzhansky decided to do so because "what stands there now could have been written by Dr. Goebbels."[51]

## DUNN

In 1946 Dobzhansky coauthored his first popular book on the race concept, *Heredity, Race, and Society*, with his Columbia colleague, the geneticist L. C. Dunn. Dunn was a distinguished geneticist who involved himself as deeply as Dobzhansky in the social aspects of scientific research. By the time of their collaboration on *Heredity, Race, and Society*, Dunn had established himself as one of the world's preeminent geneticists, particularly in the area of mouse genetics and gene mapping, and is considered to have played a crucial role in the establishment of genetics as its own scientific discipline. In that capacity Dunn served on the Joint Genetics Section of the American Association for the Advancement of Science during the

1920s and early 1930s, as the first president of the Genetics Society of America in the early 1930s, and also helped to transition the *Journal of Heredity* from a eugenic mouthpiece into an academic journal.[52]

Dunn also facilitated the emergence of the modern synthesis through his relationship with Dobzhansky. The two geneticists had known each other since at least the mid-1930s, when, at the invitation of Dunn, Dobzhansky delivered a series of lectures at Columbia in 1936 (and a year later was named as part of the revitalized Jessup Lecture Series at Columbia).[53] Those lectures, at the prompting of Dunn, were adapted into *Genetics and the Origin of Species*.[54] After reading the book, Dunn acknowledged its importance, writing to Dobzhansky that "it fills a need we have all felt for a long time."[55] In 1940 Dunn would also be responsible for bringing Dobzhansky to Columbia from Caltech, where he had been working with T. H. Morgan since the late 1920s. Thus began a lifelong friendship and collaboration that lasted until Dunn's death in 1974.[56]

Dunn's interest in race and genetics preceded his collaboration with Dobzhansky on the subject by more than twenty years. He was, in the early 1920s, active in the American eugenics movement, even delivering a paper at the "Second International Congress of Eugenics" held at the American Museum of Natural History in 1921. During the 1920s Dunn published several papers on race and race mixture that indicated his growing distance from the eugenics movement, including his 1925 paper "A Biological View of Race Mixture," which argued "popular assumptions of hybrid inferiority are shown to lack support. Biological evidence indicates that neither inbreeding nor outbreeding has uniform effects, and that each case of crossing may have to be considered as a special problem."[57] Dunn's conclusions ran counter to most eugenic and genetic work being done at the time, and, like Dobzhansky, he seemed to share a desire to both explain the race concept in the context of the modern synthesis and to depoliticize and preserve it for scientific purposes. In his 1928 study called "An Anthropometric Study of Hawaiians of Pure and Mixed Blood," part of the research originally presented at the American Museum of Natural History in 1921, Dunn tried to make this point. The paper argues both that there was currently not enough good data (which Dunn himself would collect) and that, based on this work, he and others could help develop a racial definition for Hawaiians.[58]

Dunn held a divided opinion of eugenics during the 1920s, according to one of his biographers. On the one hand he accepted "biological eugenics" as a valid extension of Mendel's laws to humans. On the other hand, he was skeptical of the application of social eugenics to human populations, recognizing early on that heredity was complex and unpredictable, and that sterilization laws could not account for such complexity. Dunn's position was complicated by his own evolving views on the subject, driven by both the changing scientific landscape and circumstances in his personal life. In 1928 Dunn's second son, Stephen, was born with cerebral palsy.[59] Although Dunn would always believe in the potential of applying genetics to social problems in human populations, by the 1930s he came to reject the eugenics movement for its mix of science and politics, and for the fear that a child like Stephen, who despite his physical handicaps grew up to receive a Ph.D. in anthropology from Columbia, would have been deemed unfit in most eugenic circles. Later that decade, Dunn would assist the Carnegie Institution of Washington—then still the principal sponsor of Charles Davenport's work at Cold Spring Harbor—in evaluating the work of the Eugenics Record Office. In a letter sent in 1935 to John Merriam, president of the Carnegie Institution, Dunn summarized his views of eugenics: "Eugenical research was not always activated by purely disinterested scientific motives, but was influenced by social and political considerations tending to bring about too rapid application of incompletely proved theses."[60] The Eugenics Record Office was officially closed in December 1939, when it had become abundantly clear that its work was scientifically unacceptable in the United States and increasingly objectionable, particularly in light of the extremes of eugenics of the Third Reich in Nazi Germany.[61]

That Dunn rejected eugenics and continued to believe that genetics could be (carefully) applied to human populations was similar in many ways to Dobzhansky's thinking on the subject, and the two men, as well as many of their genetics colleagues, saw no contradiction in rejecting what they considered a politicized eugenics movement and at least theoretically embracing its possibilities. In the 1950s Dunn even founded the Institute for the Study of Human Variation at Columbia, an effort to identify the "causes of evolutionary changes in human populations" and to see how such evolutionary changes "could be initiated and developed."[62]

Dunn's eugenic venture was short-lived. The institute didn't survive the decade, and Dunn retired from his position at Columbia in 1962. Dunn's work through the institute reflected his long-standing interest in issues of discrimination and equality as they intersected with science. He launched two population genetic studies, one on the Jewish population of Rome and one on the Gullah peoples of James Island, South Carolina. Although his scientific hypotheses sought to examine the evolution of human populations, Dunn's work on these groups also addressed issues of ongoing discrimination, particularly the role of discriminatory social isolation as a factor in human evolution.[63] Dunn's student Dorothea Bennett remembered his devotion to the potential of genetics in this way: "He was equally interested in man, mouse and garden flowers in the sense that they presented interesting biological correlates that were just as real to him although their overall importance might be very different."[64]

At the peak of their influence in the 1940s—Dobzhansky because of his role in the synthesis and Dunn because of his role in establishing the discipline of genetics—they were well positioned to author a popular book on genetics, race, and racism. And their collaboration on *Heredity, Race, and Society* bore unquestionable fruit; its print run exceeded 500,000 copies over its four editions, and it was translated into many languages as well. The book brought together their personal and disciplinary views on race, racism, and genetics: that race was a useful biological concept that, in itself, did not buoy racism. Instead, racists misused genetics for their own nonscientific ends. The book also ventured into the nature-nurture debate, arguing that "between so-called 'hereditary' and so-called 'environment' traits there is no hard and fast line."[65] Finally, Dunn and Dobzhansky acknowledged the historical nature of the race concept, writing, "The belief that the differences between men and races are inborn and unalterable is probably older and more widespread than the other extreme view that human peculiarities result from peculiarities of the environment in which they occur."[66]

The book, directed at a general audience, meant to "acquaint them with the biological facts necessary for understanding human likenesses and differences."[67] In its chapters titled "Human Differences," "Nature and Nurture," "The Method of Heredity," and "Group Differences and Group Heredity," Dunn and Dobzhansky updated their readers on the state of genetics and evolutionary biology, the failings of eugenics, and the emergence of population thinking. But it is their final chapter on race where they sought to leave a mark on general science readers, or on "John Q. Busrider," whom a derisive reviewer suggested was the target audience of their book.[68]

In the book's final chapter, "Race," Dobzhansky and Dunn made a case for why race is an important concept for science and society, and why racism is not supported by science. These arguments were meant to support their belief that the race concept could and should be divorced from racism and racists. "Races can be defined as populations which differ in the frequencies of some gene or genes," they wrote, using scientific language to describe the nature of differences between human groups. And they emphasized the point that even though "military leaders and politicians have been learning how to use real or assumed scientific discoveries to add an appearance of respectability to their propaganda," that propaganda "cannot accomplish its purpose if we know the facts of human biology."[69] In other words, even the biological race concept endorsed by Dunn and Dobzhansky could provide support and cover for racist thinking, since science could also challenge that use.

Yet reading the book, one can't help but wonder if Dunn and Dobzhansky were aware of the contradictions in their reasoning about the nature and meaning of the race concept and why they believed it to be such an important classificatory tool for biology. The book's final chapter is marked by contradictions obvious to any observer. On the one hand, the authors assert "races are populations which differ in the relative commonness of some of their genes," yet at the same time they write tentatively that "when we say that populations are racially different we are not saying very much. They may be so different that it is possible to tell to which of them any individual belongs, or so similar that only very careful study by specialists can reveal their distinctions at all."[70] This position seems to assert that ultimately only "specialists" can determine race, and that what laypersons may think they know about race is, in fact, inaccurate. Similarly, Dunn and Dobzhansky wrote, "Human races differ usually in many genes and many traits," and also stated that "attempts to subdivide mankind neatly into several hard and fast racial compartments evidently failed."[71] Moreover, Dunn and Dobzhansky believed that the existence of human races would eventually cease to exist, that "there is no doubt that civilization leads slowly but inexorably, toward breakdown of the race divisions."[72]

The book's reception was decidedly mixed. In the *Quarterly Review of Biology* Bentley Glass recognized the book as "outstanding for the clarity and simplicity of its exposition," and wrote that "the authors have not been afraid to express their own view, growing out of their understanding of population genetics, on such subjects as the difficulties of eugenics, evolution, and 'social Darwinism.'"[73] Other reviews were not as kind. Writing in the *Journal of Heredity* Robert Cook and Jay Lush worried that the book might leave its readers' "thinking about individual and racial differences dangerously muddled." Cook and Lush also accused Dunn and Dobzhansky of "ideological biases and value judgments" and hinted that the authors harbored communist sympathies.[74] An unsolicited review sent directly to Dobzhansky by Frederick Osborn, president of the American Eugenics Society and a founder of the Pioneer Fund, attacked the book's positions on eugenics and race. "As you know," Osborn began, "I felt very badly that in the little Penguin book published by yourself and Dr. Dunn, you failed to recognize the authoritative statements on eugenics and leveled your fire instead at social class eugenics and racism, which were discarded by all responsible people many years ago." Dobzhansky mocked Osborn's critique in a letter to the anthropologist Ashley Montagu: "Ho-Hum!!! I really did not know that he felt badly, although I sort of guessed he might. And when, at what date, have 'all responsible people' discarded 'social class eugenics and racism'? I wish he might, as President of the Eugenics Society, make the discarding quite explicit and official."[75]

Dunn and Dobzhansky's postsynthesis approach to race was not the only position on race being offered in public and scientific circles. In 1948, Harvard University Press published *Human Ancestry: From a Genetical Point of View*, by R. Ruggles Gates. The book was a mix of eugenic and polygenic thinking, and Gates was an advocate of the position that races did not exist because the so-called races of man were actually different species, among which a hierarchy of abilities existed. In correspondence with Ashley Montagu, who had reviewed Gates's book for the *Saturday Review*, Dobzhansky refers to *Human Ancestry* as "excrement." But Dobzhansky, ever the arbiter of the misuse of heredity and race by scientists and nonscientists alike, was less concerned with Gates's opinion than he was with the work of the evolutionary biologist C. D. Darlington, a pioneer in chromosome studies. Darlington believed that not only did races and classes of humans have different values but also that the languages spoken around the globe were genotypically controlled. Because Darlington was no scientific kook or fringe eugenicist, Dobzhansky worried that his theory of race had "dangerous potentialities of spreading its evil smell" and feared in its wake "a big comeback of racialism in a virulent form!"[76]

For Dunn and Dobzhansky the biological race concept itself was not a problem, and they believed that they had written an important book that made that point. "As far as spreading the light is concerned this is what we (or I at any rate) owed to members of our species," Dobzhansky wrote to Dunn in November of 1946, just after the book's release. "How much light will come from our efforts is another matter and one not under our control," Dobzhansky continued. "At any rate we tried honestly."[77] That "honestly" for the two geneticists was that in the context of the modern synthesis, race made perfect sense and the public needed to understand what that meant. It was the social application of this idea that caused trouble. As such, racism, not race, was what scientists and others needed to be vigilant about. They sought to undermine scientific racism, not the biological race concept. The problem with this argument, of course, is that they simultaneously argued that race was imprecise and arbitrary as a biological category (outlined above as the "difficulties" of using race) and that it was a legitimate scientific concept. On the face of it, this is not necessarily contradictory. Biological concepts can be imprecise and still have utility. The evolutionary biology term "homology," which shares with "race" a similar imprecision and a similar struggle among scientists to define its precise meaning, has been the subject of significant debate within biology.[78] However, unlike "homology," "race" has both social and scientific meanings.

## MONTAGU

In 1942 the British-born anthropologist Ashley Montagu wrote *Man's Most Dangerous Myth: The Fallacy of Race*, the first book by a natural scientist to call for the abandonment of the biological race concept.[79] Montagu was a controversial figure who, throughout his long career (he was a prolific contributor to anthropological, biological, and sociological thought well into his nineties), took on controversial subjects with aplomb. Montagu trained first in England in social anthropology at the London School of Economics with Bronislaw Malinowski and later in the United States in cultural anthropology at Columbia University under Franz Boas and Ruth Benedict. Following the completion of his Ph.D. at Columbia, Montagu taught anatomy at Hahnemann Medical College in Philadelphia and in 1949 founded the Anthropology Department at Rutgers University. Montagu's position at Rutgers was short-lived, however. A staunch antiracist, integrationist, and early feminist, as well as an outspoken critic of the rabid anticommunist senator Joseph McCarthy, Montagu was driven from his position at Rutgers in 1953.[80] Montagu's expertise crossed emerging anthropological subfields and he published widely in areas that today would be considered both cultural and physical anthropology. His obituary in the *American Anthropologist* referred to him as an evolutionary anthropologist.

Beginning in the late 1930s Montagu began his assault on the race concept. At first, however, he only articulated an opposition to racism. In 1939 Montagu presented a motion at the American Association of Physical Anthropologists stating that physical anthropology provided no scientific basis for discrimination based on race, religion, or linguistic heritage. The motion was not passed.[81] By the 1940s Montagu began a direct attack on the race concept itself. Montagu's struggles with the race concept were similar to Dobzhansky's and Dunn's and were drawn largely from Dobzhansky's work *Genetics and the Origin of Species*. But whereas Dobzhansky and Dunn rejected a typological approach to race in lieu of a genetical approach, Montagu would reject both.

In a paper presented at the April 1940 meeting of the American Association of Physical Anthropologists (published in 1942 in *American Anthropologist*), Montagu attacked the anthropological/typological race concept, writing, "What a 'race' is no one exactly seems to know, but everyone is most anxious to tell." Instead of upholding an outdated and scientifically useless typological approach to race, Montagu argued that an anthropologist should investigate human diversity "not as a taxonomist but as a geneticist, since the variety which is loosely called 'race' is a process which can only be accurately described in terms of the frequencies with which individual genes occur in groups which represent adequate ecologic isolates." The influence of Dobzhansky's work is clear in Montagu's when he writes that "race formation is genetically best understood in terms of the frequency with which certain genes become differentiated in different groups derived from an originally relatively homogeneous species population and subsequently undergo independent development."[82]

But it is at this point—where Montagu accepts the populationist view of race—that he departs from the growing consensus on this matter. Instead of trying to save the race concept for the biological sciences (including physical and evolutionary anthropology), he rejects it out of hand, questioning, "What aggregation, then, of gene likenesses and differences constitutes a race or ethic group?" Montagu believed that the race concept, even with Dobzhansky and Dunn's modifications, was a convoluted idea, "a rather fatuous kind of abstraction, a form of extrapolation for which there can be little place in scientific thought."[83] Alternatively, Montagu proposed substituting the term "ethnic group" for "race." This was not, in Montagu's reckoning, simply a terminological sleight of hand. Montagu's notion of an "ethnic group" was a rejection of the static and immutable notion of race, a denial of "the unwarranted assumption that there exist any hard and fast genetic boundaries between any groups of mankind," and an acceptance of the genetic unity of humanity. Up to this point, his approach was no different than Dobzhansky and Dunn's. Montagu distinguished his position by his belief that even a revised race concept could not transcend the idea's long and disconcerting history in the natural sciences, especially anthropology. In Montagu's estimation, the history of physical anthropology from the middle of the nineteenth century was "a gradual inversion of this genetic approach to the problem of the variety of mankind. The investigation of causes steadily gave way to the description of effects."[84]

Montagu wrote to Dobzhansky explaining why he believed "ethnic group" was preferable to "race." But Dobzhansky, with whom Montagu carried on a friendly forty-year correspondence beginning in the 1940s, made clear his opposition to eliminating race from the biological parlance.[85] Dobzhansky was not alone in his criticism of Montagu's proposed abandonment of race; Montagu's anthropological colleagues were also quite critical. At a 1941 celebration on the occasion of the fiftieth anniversary of the founding of the University of Chicago, Montagu read a paper titled "The Meaninglessness of the Anthropological Conception of Race." The anthropologists in attendance were not supportive. From Fay-Cooper Cole to Harry Shapiro to Aleš Hrdlička, the rejection of Montagu's idea was overwhelming.[86] Resistance to Montagu's position was driven by the novelty of his ideas about race and ethnicity in the face of a field whose work was defined, in large part, by its embrace of a biological conception of race, and by both the covert and overt racists who popularized the discipline.[87]

Montagu believed there were several reasons why "ethnic group" was a better term than "race." First, he argued that "it emphasizes that we are dealing with a distinguishable group" that has "been subject to cultural influences." Second, the term "is non-committal, and leaves the whole question of the precise status of the group on physical or other ground open to question." Third, and finally, it eliminated what Montagu called "any obfuscating emotional implications." For Montagu, "the term ethnic group is not merely a substitute . . . it's a new concept, a concept for human groupings which modern knowledge has for the first time made possible."[88]

In 1942 Montagu expanded his ideas about abandoning the race concept in the book *Man's Most Dangerous Myth: The Fallacy of Race*. The book, written more than seventy years ago, remains in print—an indication of its lasting significance and impact. At the time, it quickly rose to become a best seller. As in his earlier publications on the race concept, *Man's Most Dangerous Myth* argued "that the term 'race' itself, as it is generally applied to man, is scientifically without justification and that as commonly used, the term corresponds to nothing in reality."[89] Despite chapters on "The Genetical Theory of 'Race'" and "The Biological Facts" (Montagu embraced Dobzhansky's views of human genetic diversity), he continued to reject the biological race concept for the same reasons he outlined in earlier articles—that the race concept was a vestige of history, that it measured differences that were "more or less temporary expressions of variations in the relative frequencies of genes," and that "ethnic groups" was a viable alternative term for describing the states of human genetic diversity.[90]

Reviews of the book reveal how novel and revolutionary the post-modern synthesis view of race was at that time, and how many remained unfamiliar with its propositions. Writing in the *Annals of the American Academy of Political and Social Science*, the Smith College psychologist Frank Hankins wrote that the book "falls somewhere in the twilight zone between strictly scientific treatment of its subject and propaganda," and clearly without knowledge of Dobzhansky's and others' work on a genetical race theory, attributed its ideas solely to Montagu. Hankins's review revealed his own typological thinking and fundamental misunderstanding of populationist thought: he wrote that "instead of trying to deny what the man in the street finds all too clear, namely, that there are such things as race differences," Montagu should instead have simply emphasized overlap between racial groups.[91]

Other reviews were more sympathetic of Montagu's project, although not always of the book's sometimes-polemic style. Writing in *Isis* the Harvard cultural anthropologist Clyde Kluckhohn agreed "with the central thesis that the traditional view of 'race' which has prevailed in anthropology is utterly inadequate in the face of contemporary genetics and experimental biology."[92] In the *Quarterly Review of Biology* the geneticist Bentley Glass acknowledged Montagu's thesis, stating, "One point should be clearly understood from the beginning by every reader. The author is not denying the existence of human ethnic groups which would fit the genetic definition of geographic race. He is careful throughout the book to make it clear that it is the older anthropological conception of 'race' that he regards as utterly fallacious and pernicious."[93]

In the foreword to the book, the writer Aldous Huxley, no stranger to genetics and its controversies—he was the brother of Julian Huxley and grandson of Thomas Henry Huxley (both distinguished biologists) as well as the author of the genetically dystopic *Brave New World*—offered a bleak assessment of the impact of *Man's Most Dangerous Myth*. Although he acknowledged the book's "great merits" in exposing the fallacies of race, he lamented its limitations, writing that educating the public about the limits of the race concept was not enough. Facts were not enough, believed Huxley, because "most ignorance is voluntary and depends upon acts of the conscious or subconscious will." Race and its associated ideas and behaviors would, for the moment, continue to win the day because "facts are mere ventriloquists' dummies, and can be made to justify any course of action that appeals to the socially conditioned passions of the individuals concerned."[94]

Despite Huxley's admonition, Montagu continued, both in scholarly and popular publications, to argue against the biological race concept. Montagu also published widely in the 1940s, seeking to debunk American's racial prejudices against African Americans.[95] Given that in the 1940s and the 1950s the crucible of race would reach fever pitch in American society, it is no surprise that Montagu, who had risen to become one of the most prolific and dynamic scientists writing about race, found himself at the center of that storm as head of a United Nations Educational, Scientific and Cultural Organization committee on the race concept.

# 8

# CONSOLIDATING THE RACE
# CONCEPT IN BIOLOGY

On May 17, 1954, the United States Supreme Court ruled in the case *Brown v. Board of Education of Topeka, Kansas.* The Court's unanimous decision struck down legal segregation in America's public schools. "In the field of public education the doctrine of 'separate-but-equal' has no place. Separate education facilities are inherently unequal," Chief Justice Earl Warren wrote in the decision.[1] This victory for justice, this blow against institutionalized racism, was a historic moment in the intensifying civil rights movement and marked its most significant victory against legalized segregation to date.[2] Debates still rage today regarding the impact of the desegregation decision on public education. Some have been critical of the Court's vague language—"with all deliberate speed"—in setting a timetable for desegregation, others have questioned the wisdom of the Court's reliance on psychosocial evidence in showing segregation's harm on black children, and still others worried that the ruling "defamed all-black schools and their teachers." It was, nevertheless, a watershed decision.[3]

*Brown* did not comment directly on the nature of race or on the alleged superiority or inferiority of racial groups. By rejecting the precedents that had shaped legalized segregation in the United States, the justices had to articulate an intellectual foundation for their ruling, and in doing so would help reshape the intellectual terrain upon which popular notions of race were built.[4] The Court turned to contemporary social scientific evidence—psychological and sociological studies documenting the deleterious effects of segregation on African American children—to buttress its decision. In citing the effects of segregation on black children and the inevitable damage that it caused, the Court made a powerful statement, albeit obliquely, on the nature of race difference. Writing for the Court, Chief Justice Warren stated "that to separate them [black children] from others of similar age and qualifications solely because of their race generates a feeling of inferiority as to their status in the community that may affect their hearts and minds in a way unlikely ever to be undone."[5] Implicit in identifying segregation's harmful impact on black children's psyches and the wrongness of causing these children "a feeling of inferiority" was an acknowledgment by the Court that perceiving differences of race as inferior and superior is erroneous. These powerful words marked a watershed in American legal thinking, simultaneously rejecting the racist underpinnings of the Court's earlier segregation sanctioning decision of *Plessy v. Ferguson* (1896), rejecting a hierarchical view of race differences, and seeking to use this worldview to protect African American children from the injuriousness of segregation and the racist thinking that brought it about.

The Court acknowledged the impact of scientific thinking on its ruling in footnote 11 of the decision, which cited, among others, the work of Kenneth B. Clark, E. Franklin Frazier, and Gunnar Myrdal.[6] The work of the social scientists cited in this footnote sought to reframe the debates on race as a moral problem and as a problem of race relations. These analyses marked an important shift away from debates about the biological basis of racial hierarchy toward a discussion about the nature of race oppression and the ways in which the hierarchical relationship between blacks and whites had circumscribed opportunities for African Americans in the United States.[7] By including these works in the ruling's footnotes, the Court was, to a great degree, endorsing their contents and theses. Although it is Kenneth and Mamie Clark's now famous "doll test"—a psychological test designed to measure the effects of segregation in children—that received the greatest attention for its impact on the Court's thinking in striking down legalized segregation, the placement of Gunnar Myrdal's *An American Dilemma: The Negro Problem and Modern Democracy* as the final citation in footnote 11 is striking, especially given the book's critical assessment of American racial oppression and its rejection of the then widely, if not popularly accepted, typological view of race and race difference.[8]

Assessments of the impact of social scientific findings on the Court's decision in *Brown* are decidedly mixed, particularly on the meaning of footnote 11. One school of thought argues that the inclusion of footnote 11 and other evidence indicates the considerable influence of social science research in finding public school segregation unconstitutional. This group of scholars suggests "the *Brown* verdict either justified or vindicated the Court's presumptive reliance on social science evidence." Others are critical of the Court's supposed reliance on social science data, worrying that it "did not provide a sufficiently enduring basis to support the principle of racial integration." These scholars worried that a ruling based on social science data "violated traditional, more resilient approaches to constitutional interpretation." Still others, from Robert Bork to Lawrence Friedman, have held that "the *Brown* Court did not rely on social science evidence in striking down school segregation." Bork, for example, has argued that "social science evidence was immaterial to the Court's decision to strike down school segregation." Finally, more recent scholarship suggests that "footnote eleven was a by-product of the Court's legitimacy concerns." In this view, the Court was concerned about both the controversy the verdict would attract and its rationale for its decision. The inclusion of footnote 11 was therefore considered post hoc and was more important to the decision's legitimacy than its formulation. All but the first argument—that footnote 11 indicates the fundamental role that social science research played in *Brown*—fail to acknowledge the significance that the research played in Thurgood Marshall's arguments before the Court and in the cases that led to *Brown*. Furthermore, even if the impact of social science research was not definitive, it was significant. The inclusion of the Myrdal study alone indicates that the Court at least understood, and to some degree embraced, those findings. It is unthinkable that the Supreme Court would have included such a controversial study without understanding its implications and meanings for America's discussions of race and race difference.[9]

Myrdal's more than 1,000-page tome, supported by the Carnegie Foundation and written with the research assistance and consultation of more than seventy-five people, sought to recast America's racial problems as a moral conflict between the egalitarian impulses of America's democratic creed and its racist practices. Myrdal believed that this moral conflict, which he cast as the American dilemma, was soon going to resolve itself on the side of America's creed, and his work anticipated many of the victories of the then nascent civil rights movement.[10] However, for Americans to rationalize this dilemma, the tenets of white supremacy—white over black in an unchangeable biological hierarchy of races—had to be an integral part of its zeitgeist. Myrdal wrote that "race dogma is nearly the only way out for a people so moralistically equalitarian, if it is not prepared to live up to its faith. A nation less fervently committed to democracy could, probably, live happily in a caste system with a somewhat less intensive belief in the biological inferiority of the subordinate group. *The need for race prejudice is, from this point of view, a need for defense on the part of the Americans against their own national Creed, against their own cherished ideals.* And race prejudice is, in this sense, a function of equalitarianism. The former a perversion of the latter."[11]

Myrdal also spent several chapters of the book explaining the emerging consensus on the race concept that was developing in the biological sciences, and acknowledging that fixed biological notions of race rooted in typology were "being replaced by quantitative notions of the relative frequency of common ancestry and differentiating traits. . . . The great variability of traits among individuals in every population group is becoming stressed, and the considerable amount of overlapping between all existing groups increasingly recognized. . . . The fundamental unity and similarity of mankind—above minor individual and group differentials—is becoming scientifically established."[12] These "quantitative notions," of course, all point back to Dobzhansky and others working under the rubric of populations and gene frequencies in the modern synthesis.

Given that the findings of *An American Dilemma* went far beyond the narrow claims of the other works cited in footnote 11 (almost all examine, in some way, segregation's psychic toll on African American children), the inclusion of Myrdal's work here seems both purposeful and remarkable. That all members of the Court embraced Myrdal's conclusions seems unlikely, especially given several justices' sympathies for the practice of segregation, even in the wake of *Brown*.[13] However, the content of *An American Dilemma* and its inclusion in the *Brown* ruling suggest that something was changing in America's understanding of what race was and how that emerging understanding of race was altering America's racial calculus.

Published in 1944, *An American Dilemma* was a sensation from its first printing. The author Frances Gaither, reviewing for the *New York Times*, called it "a book which nobody who tries to face the Negro problem with any honesty can afford to miss."[14] Writing in the journal *Phylon*, W. E. B. Du Bois wrote that the book was "monumental," that "Myrdal does not gag facts," and that the book "does not appease the South."[15] The historian Oscar Handlin, who reviewed the book on its twentieth anniversary, observed that *An American Dilemma* was "a magnet to scholars and a catalyst to political groups," and that "its recommendations have helped shape the strategy of every organization interested in legislation and in judicial interpretations."[16]

Others were not so kind. In the wake of *Brown*, the historian Walter Jackson noted "southern members of Congress and right-wing activists in other parts of the country regularly portrayed *An American Dilemma* as a Communist-inspired work." Federal Bureau of Investigation chief J. Edgar Hoover even ordered an investigation into the book, Myrdal himself, and his many researchers. An FBI list accused forty-one of the individuals cited by Myrdal in his preface as having been or currently "members of the CP [Communist Party], CP sympathizers or members of front organizations."[17]

*Brown* heralded a shifting American racial zeitgeist that was fueled in part by changes in natural and social scientific thinking about race. Both scientific and public debates on this subject, not surprisingly, intensified in the 1950s as racists feared what the ruling and other challenges to America's racial hierarchy would bring. Mississippi senator Theodore Bilbo had warned in 1947, for example, that the mingling of races would lead to the "undermining of both the white and Negro races in this Nation."[18] Wesley Critz George, an anatomist at the University of North Carolina School of Medicine believed segregation necessary. In its absence the United States would "sacrifice our children on the altar of integration."[19]

## SO LONG RACE . . . IN A FEW HUNDRED YEARS

Just four months after the *Brown* decision Curt Stern, a renowned expert on *Drosophila* and a human geneticist at the University of California, published a controversial article about race under the title "The Biology of the

Negro" in the popular science magazine *Scientific American*. The article was a popularization of a scientific paper on the same subject that Stern had published just a year before.[20] A German-born biologist, Stern, like Dobzhansky, came to Morgan's fly room at Columbia on a Rockefeller Foundation scholarship in the 1920s (Stern left a year before Dobzhansky arrived). Based on his work in Morgan's lab, Stern produced several pioneering papers on the chromosomal basis of inheritance by looking at sex-linked chromosomal abnormalities in *Drosophila*. Stern's early work was fundamental to the development of the chromosomal theory of inheritance. Stern returned to Germany for several years following his work at Columbia, coming back to the United States to work with Morgan—again on a Rockefeller scholarship—in 1932. Hitler's rise to power a year later left Stern, a German Jew, with little choice but to stay in the United States. He spent most of the 1940s as the chairman of the Department of Zoology and the chair of the Division of Biological Sciences at the University of Rochester. In 1947 he joined the zoology faculty at the University of California, Berkeley, where he would remain until his retirement in 1970.[21]

In addition to his work on chromosomal inheritance, Stern would also make significant contributions through his work on somatic cell crossing over in *Drosophila*, his development of the concept of isoalleles (different forms of a gene that produce the same phenotype or very similar phenotypes),[22] and his use of genetic mosaics to understand development in *Drosophila*. Stern also made an important contribution to radiation science. During World War II, Stern's group at Rochester exposed *Drosophila* to low levels of radiation, concluding that "there is no threshold below which radiation fails to induce mutations."[23] Perhaps Stern's greatest impact on the field was through his textbook *The Principles of Genetics*. First published in 1949, it quickly became the classic genetics primer for undergraduate and graduate students, selling more than 60,000 copies and appearing in three editions during his lifetime.[24]

Unlike others in the field whose participation in debates about human race differences were largely theoretical—neither Dobzhansky or Dunn, for example, had much research experience in the area of human genetics—Stern, beginning in the late 1930s, became increasingly involved in human genetics research, especially through teaching and training doctoral students. He published widely and lectured on the inheritance of skin color and on sex-linked inheritance and was, in 1957, the president of the American Society of Human Genetics. But Stern's forays into race and genetics were not without controversy. A February 1946 lecture at the University of Rochester titled "Why Do People Differ?" stressed "the unscientific attitude of any race prejudice." The program for the lecture pictured a cartoon with three girls, two of whom were white, one black. The local chapter of the NAACP objected to the picture of a shabbily dressed black girl in pigtails wearing boots flanked by two well-dressed white girls looking at the black girl and smiling. Stern defended himself and his lecture, claiming that "the purpose of the cartoon was to indicate differences of heredity (color) and of custom (different kinds of dresses)."[25]

Stern was well established in human genetics by the time that "The Biology of the Negro" was published in *Scientific American*. The magazine, which had held up publication of the article because "of the priority of other subjects which seemed more immediately timely," realized that in the wake of *Brown* they had an article on their hands that would generate significant readership. In August of 1954 Leon Svirsky, the managing editor of *Scientific American*, wrote to Stern, telling him that "now seems an excellent time to print an article on the Negro in view of the recent decision on schooling."[26]

The article confirmed the worst fears of ardent segregationists. According to Stern, at the time he wrote his article almost 80 percent of "U.S. Negroes" had white ancestry; by 1980, he believed, "there will hardly be a single Negro in the U.S. who can claim a purely African descent."[27] Furthermore, Stern predicted that the future would witness only an increase in the "flow of African genes into the numerically dominant white population," that the average skin color of the American population would "shift slightly toward a light brunette," and that "complete fusion" between American whites and blacks would leave only a "few thousand black people in each generation in the entire country." If someone living in 1954 "could return at a distant time, he would ask in wonder: 'What became of the Negro?' "[28] In Stern's estimation, not only would the "Negro" race largely cease to exist as the result of being hybridized into the dominant white population but also race itself would become meaningless as the white and black races amalgamated. This idea infuriated racists and segregationists.

"The Biology of the Negro" impacted far beyond *Scientific American's* usual readership, eliciting intense reactions from readers and getting coverage in both national and local newspapers across the country. Personal reactions to the article were, in the racial climate of 1954, predictably swift and angry. One anonymous letter read, "You'r [*sic*] nuts—the people of America will not absorb the nasty, vile, vicious, horrible niggers. They are destroying parts of America."[29] A letter from one C. L. Barnett of Shreveport, Louisiana, to Stern in 1955 was addressed to "Old Banana Nose," an anti-Semitic reference to Stern's Jewish heritage. "That damn propaganda article you wrote about the mixing white and negros [*sic*] in the future is just about what one could expect from a Christ killing Jew. That is exactly what you Heebs want—to weaken the white man so you can take over. What is going to happen is that you and your brother smockleburgers are going to be more hated and dispised [*sic*] than you ever were in Germany." A Mrs. John Lansdell Howerton of Greensboro, North Carolina, wrote to Stern, calling to question the legitimacy of his conclusion. Howerton admonished Stern, telling him that race mixing might occur in California or the North, but that it didn't happen in the South. "The Negroes have pride in their race," Howerton wrote. "We like our Southern Negroes and they like us—we understand them and help them. . . . We do not accept the Negro socially and never will and as for intermarriage that is out of the question. . . . Did it occur to you that the Negro prefers being black? Just a little food for thought."[30]

Finally, a letter from a fictitious Marcus Julius Frogstein, M.D., in October 1954 exemplified the racist response to Stern, the fear that his prediction generated, as well as the not-so-subtle threats that Stern faced following the article's publication. Frogstein's letter was in reaction to an article in the *Fresno Bee* on September 21, 1954, about Stern's *Scientific American* article. Scribbled across the copy of the letter in Stern's papers was a note that said "not in Fresno phone book, not in AMA list." One can assume that the name was an anti-Semitic play on Stern's Jewish heritage. "Taxpayers should see that any SCIENTIST of your caliber be kicked out of U.C.," claimed Frogstein. "A drop of Negro blood is like a drop of ink. The ink will color a whole glass of water, and the COLOR AND STINK of a Negro, cousin to the baboon rather than the human race, will corrupt the blood of the White forever and forever."

Frogstein's letter also feared Stern's "philosophy" meant "extinction of the White Race."[31]

Not all responses to "The Biology of the Negro" were so negative, nor were all southerners so repelled by Stern's prediction. Several southern academics wrote to Stern, complimenting him for his thoughts on America's racial terrain. H. J. Romm, from the Department of Biology at Southern University in Baton Rouge, Louisiana, commended Stern on his piece, writing that he wished "that this article could be read by the Caucasians in the South. . . . It is not uncommon now to see articles in the paper saying that if the blacks and whites hybridize in the South a mongrel race will develop. I doubt if these people know anything about genetics." A similarly positive letter from C. H. Arndt, a botanist at Clemson College in South Carolina, thought that the article would "go a long way in dispelling some ideas that are held by individuals who have no conception of the biological factors involved in racial problems."[32]

Notwithstanding the ways in which Stern's bold predictions about the future of race in America resonated in various ways with some in the American public, the article offered important insight into the often contradictory thinking about race among scientists. Stern's piece and the reaction to it also confirm what is so obvious with hindsight: thinking about race, even by the most distinguished biologists, was often contradictory. On the one hand, in his prediction that the "Negro" race, as it hybridized into the dominant white American population, would largely cease to exist, Stern acknowledged the dynamic nature of racial categories. He also asserted, contrary to popular opinion, that "there seem to be no inherent biological weaknesses in the Negro which place him at a disadvantage" and identified "socio-economic factors" as responsible for health disparities between white and black Americans. Stern also dismissed claims of African American intellectual inferiority. On the other hand, Stern described the black race hybridizing into the white race but made no mention of the reverse being part of his prediction as well, even though this is implicit in the hybridization of the two populations. Did he not discuss this simply because it was obvious to him and his readers, or because it would have further fanned the flames? Stern's muddled thinking went further than this. He could at one moment describe Africans as the most genetically heterogeneous of any human population and in another revert to typological thinking by saying that African Americans with no Caucasian ancestry "should have both dark skin and thick lips," the assumption being that the purer an individual, the more likely to have such stereotypical African traits.[33] The stain of American society's thinking about race was never far removed from scientific thought.

## WHOSE SCIENCE DEFINES RACE?

Stern's "The Biology of the Negro" was neither the first nor the last publication on race during this time that would garner significant public and scientific attention. In 1949 the United Nations Educational, Scientific and Cultural Organization (UNESCO) convened an expert panel on race to "collect scientific materials concerning problems of race," "to give wide diffusion to the scientific information collected," and "to prepare an education campaign based on this information."[34] The founders of the program believed, as had many antiracists before them, from Du Bois to Dobzhansky, that scientific knowledge, particularly in the wake of Nazi atrocities, could be marshaled in the fight against racial prejudice.[35] The UNESCO panel on race was not the first time scientists would assemble to discuss this topic. International Eugenics Congresses in 1912 (London) and in 1921 and 1932 (at the American Museum of Natural History, New York) did much the same, although they certainly did so with a different academic and social impulse.[36] In the 1920s the National Academy of Sciences convened several committees examining race in various scientific contexts. And in 1937, the anthropologist Franz Boas and the eugenicist Frederick Osborn proposed holding an international meeting of leading scientists under the auspices of "some government or the League of Nations" to discuss "the general problem of the characteristics of race."[37] World War II intervened and the meeting was never held, but the seed was planted among international scholars to come together in some way to explore the meaning of race.

To amass, analyze, and communicate the current state of racial thought, UNESCO assembled a committee of ten esteemed anthropologists, psychologists, and education specialists. The participants came from across the globe and included such venerated scholars as the Mexican anthropologist Juan Comas, the American sociologist E. Franklin Frazier, the French anthropologist Claude Lévi-Strauss, and the American anthropologist Ashley Montagu, who was chosen as rapporteur and editor of the committee. In December 1949 the committee met over several days at UNESCO headquarters in Paris and developed its statement on race. In a later interview, Montagu recalled the committee's proceedings as fairly trouble-free, and that on the first day of the committee's meeting those in attendance each shared their views of the race concept. According to Montagu, after he shared his ideas on race, one of the committee members, the anthropologist Ernest Beaglehole from New Zealand, said, "Well, I think that Ashley Montagu has stated pretty much the view to which we would all subscribe. Why don't we get him to write this down and use it as a whetstone upon which to sharpen our wits." The following day Montagu presented the committee with his outline of the statement, and "during the following days [it was] thoroughly debated and revised."[38] In addition to distribution to committee members, drafts of the statement were distributed for comments to geneticists, biologists, social psychologists, sociologists, economists, anthropologists, and experts in labor-management relations. These experts included Dobzhansky, Dunn, Stern, the British biologist Julian Huxley, the economist and sociologist Gunnar Myrdal, and the American geneticist H. J. Muller.[39]

Published in July 1950, the UNESCO statement on race was arranged in fifteen numbered paragraphs, each making a specific point about the nature of the race concept. From recognizing the scientific acceptance of the unity of mankind to explicating the postevolutionary synthesis biological race concept, the statement began by offering a straightforward assessment of the current state of academic thinking on race. Embracing a postsynthesis definition of race, it described the biological race concept in language almost identical to Dobzhansky's: "From the biological standpoint, the species *Homo sapiens* is made up of a number of populations, each one of which differs from the others in the frequency of one or more genes." A race, then, simply defined, was "one of the group of populations" making up humankind. These "are the scientific facts," affirmed the statement. Yet the statement recognized a gap between scientific and popular thinking on race; that "when most people use the term race they do not do so in the sense above defined. To most people, a race is any group of a people whom they choose to describe as a race." As an antidote to the public's willy-nilly use of the term "race," the statement proposed "to drop the term 'race' altogether and speak of ethnic groups" instead.[40]

The statement also proposed the organization of "present-day mankind into three major divisions": Mongoloid, Negroid, and Caucasoid. While such a classification scheme had echoes of past racial doctrines (e.g., Blumenbach, Linnaeus, Buffon), the statement clearly hoped to distance itself from scientific racists in its argument that human divisions were "dynamic, not static," and that "there is every reason to believe that they will change in the future."[41] The remaining seven points of the statement sought to debunk popular confusion about race, including the assertions that mental characteristics are not included in anthropological classification of humans, that scientific evidence did not support "the conclusion that inherited genetic differences are a major factor in producing the differences between the cultures and cultural achievements of different peoples or groups, that no inborn differences exist between human groups, and that race mixture produces no physical or mental disharmonies." The influence of Dunn and Dobzhansky in the document is obvious in the fourteenth point of the statement, insofar as the document is highly critical of the popular conception of race while acknowledging its biological significance: "The biological fact of race and the myth of 'race' should be distinguished. For all practical social purposes 'race' is not so much a biological phenomenon as a social myth." Finally, the statement concluded with a summary of its pronouncements.[42]

The statement received prominent and generally positive coverage in major media outlets. A front-page headline in the July 18 *New York Times* declared, "No Scientific Basis for Race Bias Found by World Panel of Experts."[43] A day later the *Times* editorialized on the UNESCO statement, writing that the committee had "performed a valuable service in making a study of the concept of race," and that its conclusion that there was no racial hierarchy in mankind was a "truth that needs popularization." The editorial also embraced the statement's call for abandoning race as a scientific concept, stating, "To eliminate 'race' as a scientific term is a step toward ending it as a myth that dictators and movements use as political instruments to gain and exercise power." Interestingly, though, the editorial made no mention of the statement's impact on the state of race relations in the United States, focusing instead on anti-Semitism and other forms of European racism.[44] Unlike the *Times*, the *Hartford Courant* did not avoid the domestic implications of the UNESCO statement, declaring that because of it, Americans could no longer "succumb to the doctrine of white supremacy." The *Courant* also endorsed the committee's call to "drop the word race from our vocabulary," and criticized the race concept itself for having "poisoned human relations long enough."[45] Not all newspaper accounts of the statement were as supportive, however.

Academic responses to the statement on race were decidedly mixed. The geneticist William Castle, with whom Montagu had a long and sometimes tumultuous relationship, sent him a warm, congratulatory note upon its publication: "Hearty congratulations on the splendid, sane, and sound statement on race which you were able to get UNESCO to adopt."[46] L. C. Dunn was also supportive of the statement. In prepublication comments sent to Montagu, Dunn thanked him for the statement "with which I still agree heartily almost thru' out," and suggested minor revisions. Dunn hoped that his suggestions could help the statement withstand "the closest scrutiny from hostile persons without having holes . . . in it."[47] Dobzhansky, who had been a reader for UNESCO of a draft of the statement, wrote Montagu in October of 1950 to offer his congratulations, and commended him for his work and for having "done a fine job in pushing them [the statements] through." Dobzhansky noted that one of his students, a Pakistani, "started translating them . . . in order to publish the translations in some journal there."[48]

Despite support from such renowned geneticists, the statement received harsh reviews by many biologists and anthropologists, particularly for its proposal to abandon the race concept. Dobzhansky had warned Montagu about the potential furor over abandoning the race concept in the course of their correspondence (dating back, on this subject, to at least 1944). In Montagu's judgment, using ethnicity instead of race "represents a clarification, not a device or subterfuge. I may be utterly wrong, but I believe that what I have done is scientifically sound and morally desirable. . . . There is a great difference between a race and an ethnic variety, and this is what I hope I may some day convince you of."[49] During the controversy over the statement, Dobzhansky reminded Montagu of their ongoing disagreement: "The main attack of course against your suggestion of abolishing the term 'race' in favor of 'ethnic group,' and you will remember, my friend, that for the last ten years I have done my damndest to convince you that this proposition will neither be accepted nor would it do any good if accepted."[50]

In *Man*, the journal of the Royal Anthropological Institute, the British primate anatomist and anthropologist Osman Hill attacked the statement, calling it "merely the misguided opinions of a particular school of anthropologists whose assertions appear to be motivated by wishful thinking." Hill, an unreconstructed typologist, was the prosector at the Zoological Society of London and would later, in 1962, become the assistant director of the Yerkes Laboratories of Primate Biology, which in 1965 became the Yerkes National Primate Research Center at Emory University.[51] Hill denied the premises of the statement on race, decrying its claims, for example, that mental capabilities were much the same between races and dismissing its assertion that racial hybridization was harmless. Hill's writing illustrates how typological thinking remained prominent in some corners of anthropological thought at the time. He wrote, for example, that "temperament and other mental differences are well known to be correlated with physical differences. I need but mention the well-known musical attributes of the Negroids and the mathematical ability of some Indian races."[52]

In a letter to Montagu about the Hill letter in *Man*, Dobzhansky wrote that Hill's thoughts revealed the man "to be an old fogey."[53] Not all opposition to the statement was couched in such transparent typological thinking. An editorial in May 1951, again in *Man*, gave UNESCO "great credit . . . for undertaking the campaign" against racialism, but overall it considered the statement "not the effective weapon which we had looked for, but a broken reed." In *Man*'s opinion, the effort was tainted by including on its panel "a small group of philosophers, historians, sociologists and others, only two of whom had any pretensions to competence in physical anthropology." The editorial was suspicious of many of the assertions in the statement, including its rejection of polygeny, its construction of race as a genetic rather than typological construct, and its rejection of the term "race" itself.[54]

Despite what Dobzhansky perceived to be largely unfounded criticism of the statement, he informed Montagu in February of 1951 of the formation of a second committee. "The protests against the UNESCO race statement of yours," Dobzhansky wrote, "have caused UNESCO to convoke another conference in Paris, on June 4th–8th, this year, consisting of physical anthropologists and geneticists." Unable to attend, Dobzhansky was concerned about the makeup of the new panel and worried that "the genetical side of the group may consist of some people (such as Darlington) who are out and out racists." "This may result," lamented Dobzhansky, in a "statement which will be pretty sad."[55] Dobzhansky was correct in his assumption that the second committee would consist of geneticists and physical anthropologists. However, Darlington was not on the committee, and Dobzhansky's concerns were mollified by the presence of Dunn as the committee's rapporteur.

Montagu's position on the race concept is often mischaracterized in the historical literature; he is portrayed as a strict environmentalist who "opposed any scientific concept of race." Likewise, there is speculation that the tenor of the statement was a consequence of his "political zealousness and his refusal to adhere to received wisdom in science."[56] This is misleading. While Montagu wanted to abandon use of the race concept in the sciences, he proposed doing so not because he rejected the biological diversity of humankind or because he himself rejected a geneticist's or anthropologist's ability to classify humans into groups or populations or what have you. Rather, he did so because he recognized that the race concept was tainted by its long and sordid history. Montagu's statement was at once a repudiation of the typological race concept and an embrace of the postsynthesis race concept as outlined by Dobzhansky and others. As such, Montagu was very much in line with genetical thinking on race. That Montagu and the statement fell under such a cloud for its assertions is indicative, perhaps, of his lack of political expedience (Dobzhansky's prediction about removing the race concept turned out to be correct) and of the fact that his views about the biological race concept were more radical than those of most geneticists and others in the natural sciences.

A second UNESCO statement on race was issued in June 1951. J. B. S. Haldane, head of the Department of Biometry at University College, London, Gunnar Dahlberg, director of the State Institute for Human Genetics and Race Biology at the University of Uppsala in Sweden, and A. E. Mourant, director of the Blood Group Reference Laboratory at the

Lister Institute in London were among the prominent scientists on the committee. Montagu remained a committee member, and, although not official committee members, Huxley and Dobzhansky "contributed to the final wording" of the statement.[57] In his introduction to the second statement, Dunn explained the need for a revised report on race, insisting (unfairly to Montagu) that "it was chiefly sociologists who gave their opinions and frame[d]" the first statement. Alfred Métraux, the head of the Department of Social Sciences at UNESCO, who called for and organized the second committee, concurred with Dunn's assessment. Métraux was hopeful that the second UNESCO statement on race would allow the committee to go forward "without exposing ourselves to criticisms on the part of scientists."[58]

In Dunn's judgment, problems with the first statement were twofold: first, "it did not carry the authority of just those groups within whose special province fall the biological problems of race, namely the physical anthropologists and geneticists," and, second, its assertions were "not supported by many authorities in these two fields." Although Dunn acknowledged that there were important differences between the two statements, in his introduction he emphasized that "there were no scientific grounds whatever for the racialist position regarding purity of race and the hierarchy of inferior and superior races to which this leads."[59] The second statement also rejected arbitrary usage of the race concept, noting "national, religious, geographical, linguistic and cultural groups do not necessarily coincide with racial groups; and the cultural traits of such groups have no demonstrated connection with racial traits."[60]

Yet despite being downplayed by Dunn and, subsequently, many scholars since, the differences in both tone and substance between the first and second statements were significant. Whereas the first statement defined the race concept in the postevolutionary synthesis vernacular of gene frequencies and populations, the second statement, to a very large degree, contradicted this approach, suggesting that race can also be understood in typological terms as an anthropological "classificatory device providing a zoological frame within which the various groups of mankind may be arranged" that "should be reserved for groups of mankind possessing well-developed and primarily heritable physical differences from other groups." Whereas the first statement proposed dropping the "term 'race' altogether" in favor of "ethnic groups," the second statement embraced the race concept in multiple forms as an anthropological tool. Dunn himself points out that "we were . . . careful to avoid saying that, because races were all variable and many of them graded into each other, therefore races did not exist. The physical anthropologists and the man in the street both know that races exist."[61] And, whereas the first statement rejected the idea that "human hybrids frequently show undesirable traits, both physically and mentally," the second statement was more circumspect, asserting that "no reliable evidence" exists "that disadvantageous effects are produced" by racial hybrids.[62]

Even after its July 1951 release, the second statement went through multiple revisions. As late as February 1952, in correspondence with Métraux, Dunn expressed his dissatisfaction with the statement, writing, "I don't like it as it stands and would prefer to see a more thorough-going revision. . . . Since you are going to circularize the amended statement, could you not at the same time send out copies of the radical revision which I made and let the members choose between them?" Dunn believed that the difficulties in drafting both statements arose out of the attempt "to justify a particular ethical position on scientific ground"—that being, in this case, the idea that all people, regardless of race, are equal. "It is our duty as scientists to make the facts as clear as possible and to relate them to the evidence," Dunn wrote to Métraux.[63] Ultimately, at Dunn's insistence, in 1952 UNESCO published the final versions of the second statement along with the criticism received by the committee.[64]

The second statement on race was not without its critics either. The geneticist R. A. Fisher, a founder of population genetics, wrote to Alfred Métraux soon after the statement's publication. Fisher questioned the statement's position on both the nature of race itself and on the relationship between race and intelligence. Fisher claimed that "available scientific knowledge provides a firm basis for believing that the groups of mankind differ in their innate capacity for intellectual and emotional development, seeing that such groups do differ undoubtedly in a very large number of their genes." He also expressed fear that "this problem is being obscured by entirely well intentioned efforts to minimize the real differences that exist."[65] Criticism also came from those who thought that the second statement ambiguously defined race. The distinguished

University of Chicago anthropologist Sherwood Washburn, who wrote to Métraux in October 1951, pointed out what he considered to be troubling and potentially dangerous contradictions in the text, and feared that "the Statement may be used for political purposes." Washburn was particularly concerned with the statement's suggestion of the existence of three major racial groups. "'Major' in what sense?" Washburn wondered. "Certainly not anatomically more distinct than others, older in time, or numerically greatest." This line of thinking—major and minor groups—was "unfortunate," and he feared that it "opens the way for superior and inferior to come back in the picture."[66]

More than twenty years after the publication of the UNESCO statements, several geneticists, in correspondence with the historian of science William Provine, looked back at the second statement and its impact on the race debate in genetics. Sewell Wright thought it was "too broadly negative as an expression of scientific thought on the subject, however desirable such a statement may be from the political and social standpoints." On the other hand, Dobzhansky believed that a majority of geneticists—"but hard to tell how great a majority"—were supportive of the second statement. As to whether or not there was a majority or minority of opinions in favor or against the statement, L. C. Dunn felt this to be incalculable. In his estimation "it would be difficult to speak of 'majority opinions' or even of geneticists as a group. They differ in age, political and social views, class position and attitude and these are likely to affect their responses on questions which they suppose to be social or political ones."[67] C. D. Darlington, who continued to publish racist claims throughout his career, belittled the second statement, calling it "merely an expression of current fashion in emotions." And Curt Stern, whose position on race grew increasingly typological over the course of his career, wrote that it "was strongly influenced by a vocal minority."[68]

The first and second UNESCO statements on race have attracted limited but important interest from historians. For Elazar Barkan, author of *The Retreat of Scientific Racism*, the first statement exemplified "environmental determinism at its peak" and completed "the reversal in the scientific credo on race" from a eugenic view of race to a cultural notion of race proffered by anthropologists that had begun in the 1920s. Indeed, in Barkan's estimation, the assertions in the first statement confirmed the retreat of scientific racism.[69] Barkan's account, however, does not acknowledge what in actuality was the retreat from the antiracist spirit of the first statement on race or how the second statement contradicted and trivialized the first. Barkan also suggested that differences between the two statements were, for the most part, minimal and came down largely to a shift to a more tentative tone. In fact, Barkan argued that the second statement "reconfirmed the lack of scientific support for race formalism."[70] He also believed that "race was losing its scientific (read: biological) credibility, but remained powerful in popular culture and as a social institute."[71]

Like Barkan, the historian Will Provine, in an *American Zoologist* essay analyzing the relationship between geneticists and the race concept, looked at the evolution of geneticists' thinking about race differences and race crossing but did not address how geneticists conceived of the race concept itself and how that concept was developing at the time. Provine points out that while the statements indicated a general consensus among geneticists that "race mixture was biologically harmless," most accepted the belief that "hereditary mental differences probably existed between human races" even though "scientific evidence for their belief was not conclusive." Finally, Provine recognized that the race statements sought to accomplish something more significant than a summary of scientific thinking on race, and they presented an argument for "equality in society for all races while holding open the possibility that there might be average differences in intelligence between races."[72] In other words, the statements tried to make an ethical argument for equality, regardless of what the theories or perceptions of race were. The second statement failed at this largely because of its contradictory and often incoherent thinking; it rejected the first statement's refutation of the race concept and it was agnostic on whether races differed in intelligence, leaving open, as Sherwood Washburn conveyed in his criticism of the second statement, the possibility of re-creating a hierarchy of races on any number of traits. Dobzhansky, who had previously worried about the makeup of the committee writing up the second statement, was, unsurprisingly, not pleased with the outcome. After a visit to UNESCO in June 1951, Dobzhansky noted that "poor Métraux is tired and nervous, and probably about ready to send all this UNESCO business to hell."

In what was a rare moment of criticism of his close friend Dunn, Dobzhansky recorded in his personal journal that "here was also Ashley Montagu, all displeased with the statement, and largely within reason. He declared that Dunn let me down by not defending my ideas about race. In this he is wrong, for my good friend Dunn simply does not himself know or understand these ideas."[73]

More recently, the historian Michelle Brattain has analyzed the UNESCO statements as an important watershed in the history of the race concept. Brattain argues that the first and second UNESCO race statements revealed race as a historical construct in two ways: first, by showing how historical "misuses" of race had "served as rationalizations for inequality," and second, by "exposing race itself as a temporal, ephemeral phenomenon dependent on the people who invested it with authority and meaning." The great irony of the UNESCO statements was that as some of those involved in their development argued that race had a fixed, scientific meaning (particularly in the second statement), they produced a document that Brattain points out "acknowledged implicitly that race had once been one thing and now it was another."[74] But Brattain's construction of the history of the race concept relies on Elazar Barkan's notion of a retreat of scientific racism, followed by, in her estimation, a resurgence in the wake of the *Brown* decision. Brattain also argues that "scientific racism" refers only to "a particular scientific tradition associated with typological and hierarchical approaches typical of nineteenth-century physical anthropology." Brattain defends this position by noting that in the 1920s and 1930s the social sciences moved toward "a greater emphasis on environment as a cause of human variation," and that during the same time in the biological sciences "scholarship shifted toward a more complex understanding of human inheritance rooted in population genetics."[75] While this is in itself true, by positing the history of scientific thought as discontinuous, Brattain does not seem to consider the contradictions and subtleties so often present in thinking about race. A simplistic typological approach to race thinking may have been pushed to the margins of scientific thought, thus resulting in a perceived retreat in scientific racism. But, as described herein, the perpetuation of the race concept in populationist thinking confirms the very contradictions and subtleties that are ever present in the relationship between scientific practice and the race concept. Scientific racism was not pushed to the margins of mainstream scientific thought. Instead, although this was certainly not their intent, Dunn, Dobzhansky, Stern, and others preserved the race concept in biological thinking by placing it firmly in the context of modern biology.

If Michelle Brattain is correct in asserting that the UNESCO statements can help in the historian's project of historicizing the race concept, then an examination of the scientific, social, and political terrain of the 1950s would suggest that something much more complex was afoot.[76] First, the statements on race offered natural scientists an opportunity—perhaps the first—to collectively respond to the biological race concept as reimagined by Dobzhansky and others in the age of the evolutionary synthesis. For those who held on to the old typological ways of thinking about human difference, to whatever degree, the statements were an occasion to offer their resistance. Their challenges are indicative of the ways in which scientists were influenced in their approach to the biological race concept by extrascientific sources. One cannot read criticism of the statements—like R. A. Fisher insisting without evidence that races "do differ undoubtedly in a very large number of their genes," or H. J. Muller claiming that "in view of the admitted existence of some physically expressed hereditary differences of a conspicuous nature . . . it would be strange if there were not also some hereditary differences"— without recognizing the impact of social prejudices on the ways questions about race were often asked and about the intense resistance to even changing the terminology of human diversity.[77]

Yet at the same time, the reaction to the statements and the production of the second statement itself also indicate the ways in which a populationist definition of race had become consolidated in the biological sciences. Despite the significant differences between the first and second statements, both embraced the genetical concept that race was measured by the frequency of one or more genes in human populations and maintained that races were dynamic, not static, ways to understand human diversity. The comments of the sixty-nine physical anthropologists and geneticists about the statements, published by UNESCO in 1952 as *The Race Concept: Results of an Inquiry*, confirm this fact. There is very little criticism of the point in the second statement regarding a biological definition of race. However, the two parts of the statement that garnered the most criticism addressed which human populations could be considered racial groups and the relationship between race and intelligence.[78]

Second, despite the broader consolidation of the race concept in biology, the scientific row that followed both statements' publication offers insight into the contradictions and problems with populationist thinking about the race concept and illustrates the ways in which racism preserved its place in scientific practice and thought. This was not, to be sure, the intent of Dobzhansky, Dunn, and most other geneticists as they set about rethinking the biology of race in the context of the evolutionary synthesis and population genetics. In fact, Dobzhansky and Dunn were committed antiracists. In the promotion of a "modern view of race founded upon the known facts and theories of heredity," Dunn, for example, believed that "the old views of fixed and absolute biological differences among the races of man, and the hierarchy of superior and inferior races founded upon this old view" would be superseded and left "without scientific justification."[79] Dobzhansky thought it was the "misuse of the word race by propagandists and bigots" that led to popular misunderstanding of the race concept.[80] Both Dobzhansky and Dunn (very much like Du Bois nearly half a century earlier) had confidence that education about what was now considered "race" would put to rest any confusion about what it had meant in the past.

But what they refused to acknowledge, and what in hindsight Montagu seems to have been most prescient about, was that in American society it is not possible to tease apart the social and scientific meanings of race—remember this is the same conclusion that Du Bois reached as his own thinking about race evolved in the course of his early career. In Montagu's view the term "race" "enshrines so many of the errors which it is desirable to remove." Even if biologists were to use the term in its so-called modern sense, the word "race" itself would still elicit meaning in the context of the old view. Race is a "trigger word," Montagu wrote, "utter it and a whole series of emotionally conditioned responses follow."[81] One need not look any further than the changes from the first to second statements to recognize the contradictions inherent in the race concept. Even as the second statement embraced essentially the same populationist race concept as the first, it also embraced a typological race concept, stating that race is an anthropological "classificatory device providing a zoological frame within which the various groups of mankind may be arranged" that "should be reserved for groups of mankind possessing well-developed and primarily heritable physical differences from other groups." Moreover, that a group of distinguished geneticists could not come to reject claims that race mixture produced physical or mental disharmonies, or that it could claim three "major" racial groupings, highlighted the contradictions inherent in the race concept and inherent in the thinking of those who thought that modernizing the concept could make it less connected to racism.

Third, these same challenges to and changes in the statements suggest something about the status of scientific authority in the mid-twentieth century. As Dunn himself believed, the first statement was a failure because "it did not carry the authority of just those groups within whose special province fall the biological problems of race, namely the physical anthropologists and geneticists." Instead "it was chiefly sociologists who gave their opinions and frame." Dunn's critique not only ignores the contributions of the anthropologists involved in the production of the first statement (it may be recalled that Montagu had expertise in both physical anthropology and evolutionary biology), but it also belittles the important role social scientists—from W. E. B Du Bois to Gunnar Myrdal—played in helping to reformulate ideas about race.[82] Such statements point to a growing dissonance between social and scientific thought, one that the British philosopher of science C. P. Snow would, in 1959, famously describe in his essay "The Two Cultures." "I believe the intellectual life of the whole of western society is increasingly being split into two polar groups," Snow wrote. He worried that a "gulf of mutual incomprehension— sometimes (particularly among the young) hostility and dislike, but most of all lack of understanding" was emerging between the sciences and other modes of thought.[83]

The authority of evolutionary biology and genetics had, by the 1950s, according to historian of science V. B. Smocovitis, "been secured by becoming officialized and institutionalized as the central component of the biological sciences." And with this authority biologists "were to act as unifiers, negotiators of the location of biology, preservers of the whole of the positivistic ordering of Enlightenment knowledge."[84] It is in this context then that Dunn's claim about the role of sociology in the first statement can also be understood as an assertion of the evolutionists' and geneticists' power. And Dunn believed he was using this power for good. "You see, it's a much more palatable view for people who come to think about race for the first time than the typological view which had been dominant, and still is in many parts of anthropology," Dunn wrote, concluding that "my own opinion is that this was a better way to counteract ideas about race prejudice." He hoped that "other views tend to die off when people have a better one to substitute."[85] The second statement is thus as much an assertion of what committee members believed about race as it was an affirmation of their authority over social scientists on this matter.

Despite biology's authority in the mid-1950s, social scientific thought also commanded significant and growing influence in mid-twentieth century America. Witness its impact on the *Brown* decision and on thinking about American race relations more generally. Social scientific thought also examined and challenged the status quo through best-selling works like David Riesman's *The Lonely Crowd*, William Whyte's *The Organization Man*, and C. Wright Mills's *The Power Elite*. Moreover, in many ways the 1950s were a period of waning dominance for the natural sciences. No longer would the authority of science go unchallenged. In many ways Montagu's position and the position of the first statement in general were precursors to what historian Ronald Walters calls the "much larger late twentieth-century project of seeking out the hidden operations of power, questioning once seemingly fixed and 'essential' categories such as race, gender, and ethnicity."[86] That the statements were part of the push and pull between the natural and social sciences is elucidated by Walters, who argues that beginning in the 1960s, soon after the statements' publication, "scientists themselves [would do] little to enhance their image as spokespersons on crucial issues," including a growing penchant for speaking out "with so much caution and nuance that their words carried little impact."[87] Although the revisions to the first statement took place ten years before the shift Walters describes, the changes between the first and second statements suggest that perhaps this shift had begun even earlier.

Challenges to scientific authority, or an acknowledgment of the growing division between scientific and social thought in the 1950s, came from sources as diverse as Hollywood films, politicians, philosophers, as well as scientists themselves. Both Hollywood and Japanese filmmakers produced movies—*The Thing from Another World* (1951), *Them* (1954), and *Godzilla* (1954), for example—that challenged the wisdom and authority of science. *Them*—referring to ants supersized following exposure to atomic radiation at a New Mexico bomb test site that then terrorize Los Angeles—and *Godzilla* both played on fears of atomic science.[88] In *The Thing from Another World*, a "Thing" lands on the North Pole and attacks a band of Air Force personnel and scientists stationed there. To some the film was an allegory about America's fears of communism. To others it was a rebuke of science and scientific decision making in an era in which scientists could be celebrated or despised. The film's lead scientist, Dr. Carrington, an atomic scientist no less, believes that the Thing is a superior life-form from which humanity can learn a great deal. Several times in the film Carrington sabotages efforts to kill the deadly creature. The doctor is saved from himself and his unyielding pursuit of knowledge only once the Thing has been defeated.[89]

Politicians and philosophers too were well aware of challenges to the authority of the natural sciences. Writing in *Science* in 1957 the Hungarian scientist and philosopher Michael Polanyi argued that science had replaced religion as "the greatest single source of error" in human thought. Although Polanyi himself believed in objective truth and in the power of science, he acknowledged that the rhetorical power of science in silencing ideas was a dangerous turn for open intellectual exchange. "Today, when any human thought can be discredited by branding it as unscientific, the power exercised by theology has passed over to science," Polanyi wrote. This condemnation of scientific arrogance was written in the hope of checking "the abuses of the scientific method . . . in the interest of other human ideals which they threaten and in the interest of science itself."[90] Vannevar Bush, who had been President Roosevelt's science adviser and had helped shaped postwar government support for scientific research, feared that in the wake of the atomic bombings in Japan, private foundations were becoming "reluctant to fund natural science research."[91]

Finally, the content of both statements must be seen in light of the nascent struggle for civil rights that was emerging at the time, as well as changes in American social and political thought as pertaining to race.

The very same week that the second statement on race was released by UNESCO the awful side effects of American racism were in plain view in Cicero, Illinois, where, on July 12, a mob of 3,000 whites sought to prevent army veteran Harvey E. Clark from moving into a rental unit in a formerly all-white apartment building. Following an intervention by the National Guard to quell the riot, a State Department official quipped that the situation was similar to early defeats in Korea.[92] Lynching and other forms of violence against African Americans rose in the immediate post–World War II years as the South sought to reassert its racist mores and send a message to returning veterans that Jim Crow remained firmly entrenched. Several black veterans were killed or maimed in such violence. Between the summer of 1945 and the end of 1946 there were at least sixty lynchings throughout the South.[93] But racism in the United States did not always manifest itself in violence and intimidation. Jim Crow remained an organizing feature of American society in the 1950s, and that meant African Americans lived in a system of segregation and exploitation that both assumed and reinforced ideas about black inferiority. Jim Crow created economic dependency and inferiority, and in 1950 "nonwhite families earned nationally 54% of the median income of white families."[94]

Contradictions and ambiguities in America's racial calculus were becoming increasingly evident at that time, particularly in the context of international relations. In the wake of World War II and at the outset of the Cold War—two wars fought in the name of American freedom and ideals—the United States "engaged in a sustained effort to tell a particular story about race and American democracy: a story of progress, a story of the triumph of good over evil, a story of U.S. moral superiority," notes Cold War historian Mary Dudziak. "The lesson of this story," she believes, was that "American democracy was a form of government that made the achievement of social justice possible, and that democratic change, however slow and gradual, was superior to dictatorial imposition. The story of race in America, used to compare democracy and communism, became an important Cold War narrative."[95] It was in this context, of course, that in 1948 President Harry Truman signed an executive order desegregating the United States military.[96] Truman, despite his Missouri upbringing and sometimes intemperate personal attitudes toward black

Americans, became an ally in the emerging movement for African American civil rights. During his tenure he also called on Congress to protect voting rights, outlaw lynching, and enact civil rights laws. He even risked his political future during the election of 1948 by supporting, in order to court African American voters, a pro–civil rights platform for the Democratic Party.[97] Truman was, of course, mindful during these early years of the Cold War of perceptions of America at home and its treatment of its African American citizens. He expressed this concern via his 1948 special address on civil rights to Congress. In a stirring speech that went far beyond anything a sitting U.S. president had uttered on the subject of race until that time, Truman challenged America, declaring, "If we wish to inspire the peoples of the world whose freedom is in jeopardy, if we wish to restore hope to those who have already lost their civil liberties, if we wish to fulfill the promise that is ours, we must correct the remaining imperfections in our democracy."[98]

So too did the success of Ralph Bunche, a United States representative to the United Nations, indicate changes in America's racial zeitgeist. In 1950 Bunche became the first African American awarded the Nobel Peace Prize for his work on behalf of the United Nations.[99] Despite Bunche's incredible success, Michael Harrington feared that the popularization of his story (and, for that matter, at almost that same time, the popularization of Jackie Robinson's story) would, for most African Americans still living under the oppression of Jim Crow, come to represent "a tragic gap between the ideal and possible."[100] Nonetheless, Charles Johnson, president of Fisk University, who was possibly influenced by the UNESCO statements, wrote in 1951 of "significant recent gains in civil rights" and posited "it seems no longer necessary in general discussion of race relations in America for anthropologists to document fundamental mental and biological equality."[101] The optimism of Johnson's view would, however, be challenged in the years ahead as biologists and anthropologists continued to struggle over the race concept.

# 9

# CHALLENGES TO THE RACE CONCEPT

The 1950 and 1951 UNESCO statements on race embodied contradictions in the race concept that would surface over and over again in American scientific thought and in the popular expression about race. Two events in the 1960s—events that have been well documented in the historical literature as prime examples of scientific racism—are also significant in the way that they embody the ambiguities of a modern biological race concept. The first, the publication in 1962 of University of Pennsylvania anthropologist Carleton Coon's *The Origin of Races*, was a direct challenge to Dobzhansky's evolutionary-synthesis-era conception of race.[1] The second event relates to claims by the educational psychologist Arthur Jensen that racial differences in intelligence, or IQ, were hereditary and genetic, and that therefore educational redress for IQ discrepancies was largely useless.[2] Jensen's work epitomized the way the modern biological race concept could easily be appropriated by racists writing under the guise of science.

The advent of high-throughput genomic technologies has allowed evolutionary biologists and anthropologists to confirm earlier conclusions drawn from the human fossil record: that *Homo sapiens* originated in Africa and fanned out across the continents. Even recent evidence from the Neanderthal genome project, which suggests that as *Homo sapiens* fanned out across the globe some interbreeding did occur between non-African *sapiens* and Neanderthals, does not contradict the "out of Africa" thesis. In fact, the Neanderthal data supports this claim, showing that most genetic variation originated in Africa and spread globally with the earliest human migrations.[3] This contrasts with the view of Coon and many of his contemporaries in anthropology and biology who accepted Africa as the birthplace of the genus *Homo* but embraced a theory that posited that *Homo sapiens* evolved multiregionally from *Homo erectus* and eventually replaced them (hence multiple human races and a theory that bore some similarity to earlier polygenic ideas).[4]

Coon, who in 1961 had been elected as the president of the American Association of Physical Anthropologists, took multiregional ideas a step further (or perhaps backward) than the anthropological consensus of the times. He repackaged them in his tome *The Origin of Races* to read like nineteenth-century polygenist ideas of separate human race creations. Coon believed that the fossil record showed exactly five main subspecies of *Homo sapiens* that had evolved from *Homo erectus* at different times in our evolutionary history. "Wherever *Homo* arose," wrote Coon, "and Africa is at present the likeliest continent, he soon dispersed, in a very primitive form, throughout the warm regions of the Old World. Three of the five human subspecies crossed the *sapiens* line elsewhere. If Africa was the cradle of mankind, it was only an indifferent kindergarten. Europe and Asia were our principal schools."[5] Coon estimated that European ancestors had evolved from *erectus* to *sapiens* first, some 200,000 years ago, whereas the Congoids, what he calls the ancestors of most modern Africans, had evolved only 40,000 years ago in that "indifferent kindergarten." In recent history, each race had "reached its own level on the evolutionary scale."[6] And to this argument Coon brought what he claimed were the latest genetic and evolutionary biological techniques. Not only did the fossil record reveal these distinctions between humankind but so too did "research in blood groups, hemoglobins, and other biochemical features." Coon judged this as "good," believing that it was "encouraging to know that biochemistry divides us into the same subspecies that we have long recognized on the basis of other criteria." Because these biochemical markers were "invisible to the naked eye," Coon thought "they are much less controversial than the latter in an increasingly race-conscious world."[7] Of that race-conscious world, Coon believed that "racial intermixture can upset the genetic as well as the social equilibrium of a group."[8]

It is surprising that in 1962, at the height of the civil rights movement's challenge to the racial hierarchy implicit in Coon's typological and polygenic assertions, *The Origin of Races* was generally well received in both academic and popular presses, even though most reviewers acknowledged that controversy was sure to dog the book. Writing in *Science*, Ernst Mayr called *The Origin of Races* a "milestone in the history of anthropology." To Mayr, the book was "something for which to be truly thankful," for despite what he considered its significant shortcomings—"as persuasive as Coon's treatment is, it does not always convince me of the validity of the lines of descent that he postulates"—the book had "an invigorating freshness that will reinforce the current revitalization of physical anthropology." Mayr acknowledged that the book "will stir up more than one controversy," yet he believed that "the basic framework of Coon's thesis is as well, or better, substantiated than various possible alternatives."[9]

Writing in *American Anthropologist* Frederick Hulse of the University of Arizona argued that "no better text for a course in Fossil Man has yet been published" and that the book was neither a "racist tract, or . . . an echo from the past." Indeed, Hulse, despite accepting that "many of the conclusions are highly speculative in nature, and leave me quite unconvinced," believed *Origins* to be "a serious attempt to arrange the fossil evidence of human evolution in a meaningful way."[10] Even a review in the journal *Phylon*, an academic journal published by Atlanta University and founded by W. E. B. Du Bois, gave the book a positive assessment. The reviewer did, however, acknowledge that racists could use concepts in the book "to argue the inferiority of some groups."[11] *Origins* was even listed as a "Reviewers' Choice" for 1962 in the *Chicago Daily Tribune*. In the *Tribune*, Wilton M. Krogman wrote that Coon's work "is an important book about our own kind" that has "scientific validity; more, it has human dignity." Krogman's review also had some not-so-subtle allusions to his own thoughts about America's contemporary racial hierarchy and how the book confirmed this view. In *Origins* "we are told who we all are . . . and we all are *put in our places* as subspecies sharing a certain amount of likeness, dividing a certain amount of unlikeness."[12]

Despite the significant questions raised in the reviews of *Origins*, the book probably would not have garnered much attention had it not been for Dobzhansky, who, in a polite gesture toward his colleague Coon, sent him an advance copy of his review of the book (that was to appear in the *Saturday Review*) with a note that read, in part, "It grieves me

tremendously that I have to contradict your way of describing your findings. For I feel that it is indeed the unfortunate language which you are
using that creates a semantic predicament of a dangerous sort." Although
Dobzhansky never specifically identified Coon's unfortunate language,
one can assume that comments that refer to Africa as the "Darkest Continent" and an "indifferent kindergarten," or Coon's conclusions about a
relationship between race and Coon's so-called evolutionary scale, were
certainly among those that raised his ire. Coon was infuriated by the
review, calling it "an anti-racist tract," and was taken aback by accusations
of "mischievously furnishing ammunition to racists." Coon even went so
far as to accuse Dobzhansky of libel. In response, Dobzhansky called for
cooler minds to prevail, suggesting to Coon that "we control our tempers
and behave in a manner becoming for scientists." Dobzhansky also told
Coon that the "extreme violence of his reaction" to the review "came as a
very painful surprise," but that despite the esteem in which he held Coon,
he could not "avoid concluding that your reaction shows that you know
that you are wrong."[13]

Dobzhansky's article was never published in the *Saturday Review*. The
magazine returned his review to him with a fifty-dollar check, which
Dobzhansky promptly sent back.[14] Instead, the review would appear in
*Scientific American* and soon thereafter be reprinted in *Current Anthropology* alongside a response from Coon. In his review Dobzhansky
attacked the scientific conclusions of *Origins* and their impact on popular discussions about race. He challenged Coon's understanding of the
biological race concept, emphasizing "race is an entity that is not clearly
defined biologically," and called into question the centerpiece of Coon's
thesis—that there were "exactly five races of *Homo erectus* and that there
are exactly five living races of *Homo sapiens*." Furthermore, Dobzhansky
chastised Coon for what he called the book's "semantic mischief," particularly Coon's description of time lines for racial evolution from *erectus* to
*sapiens*. "A scientist should not and cannot eschew studies on the racial
differentiation of mankind, or examine all possible hypotheses about it,
for fear that his work will be misused," Dobzhansky wrote. However, he
believed academics, particularly scientists writing about issues as potentially explosive as race, needed to be sensitive to the way in which these
findings were communicated. In Dobzhansky's view, Coon had failed at

this, giving cover to those who "have repeatedly sought the support of bogus 'science.'"[15]

The anger Coon had exhibited in his personal correspondence with Dobzhansky spilled over into his published response in *Current Anthropology*. Instead of responding directly to Dobzhansky's criticisms of *Origins*, Coon challenged Dobzhansky's credentials as a scientist, claimed expertise over him in the area of physical anthropology, called him "professionally incompetent," and even privately suggested that he did "not understand the mechanisms of evolution."[16] Coon's response concluded that unlike fruit flies (Dobzhansky's subject of study), "human beings do not mate at random, but are kept apart to a large extent and quite effectively by cultural barriers such as language, religions, and such other customs as feelings about integration and segregation."[17] It seems a poor choice of words by Coon to conclude that "feelings about integration and segregation" were part of why humans did not mate at random, especially given the timing of the book's publication during the heyday of the civil rights movement's fight against segregation, and given Dobzhansky's criticism about the relationship between the scientific and the social realms on matters of race. In hindsight it is surprising that such a statement did not arouse more suspicion than it did at the time—save Dobzhansky's and that of a few other critics—about Coon's conscious or unconscious motivations for his conclusions in *Origins*.

That Coon was a cousin to the mid-twentieth century racist propagandizer Carleton Putnam—whose book *Race and Reason* Dobzhansky had just a year before called the "bogus 'science' of race prejudice" in the pages of the *Journal of Heredity*—was largely unknown at the time. Coon, as it turns out, offered Putnam significant personal assistance in the development of *Race and Reason*, and the two developed a strong bond over their shared dislike for what they believed was the dominance of Boasian equalitarianism in anthropological thought. Coon's intellectual support of his cousin helped Putnam, in turn, exploit Coon's work in his political mission. This quid pro quo worked well for both men.[18]

Dobzhansky was quite disturbed that his evolutionary biological colleagues either embraced or remained agnostic on Coon's book and their public disagreement. In early November 1962, Mayr wrote to Dobzhansky that when he read *Origins* he "saw none of the implications which you seem to see," and he even went so far as to warn Dobzhansky that he should make sure to cite the page numbers of what he perceived to be the objectionable passages in the book. If not, Mayr warned, "you will surely expose yourself to the criticism that you have accused Coon unfairly and unjustly."[19] Dobzhansky crafted a lengthy response to Mayr, G. G. Simpson, and William L. Strauss Jr., all of whom had taken issue with Dobzhansky's criticisms of Coon. In a "dear friends" letter to colleagues of whom he said "there are no other three people in the world whose opinions on evolutionary matters I would value more highly," Dobzhansky admits that their opinions of Coon's book "has caused me to do some soul searching and some re-examination of the book. But as a result, I feel obliged to stick to my guns, however much I would have preferred to have my guns stacked together with your guns." Dobzhansky told his colleagues that he considered *Origins* as fitting "the requirements of Putnam & Co," and wondered, without then knowing the full relationship between Coon and Putnam, whether Coon had framed his conclusions "deliberately to give ammunition to that gang."[20]

In many ways Dobzhansky's review of *Origins* and the related correspondence read as if Dobzhansky saw himself as the caretaker of the biological race concept. He had, after all, helped usher in an era in the evolutionary sciences that sought to detach the idea of race from its typological origins and, in so doing, had preserved a concept of human difference for geneticists to utilize in their work. However, despite a nearly career-long fight to preserve the biological race concept, Dobzhansky began to express severe misgivings about the concept at around this time. In his book *Mankind Evolving*, published the same year as Coon's *The Origin of Races*, he worried that investigation into human diversity had "floundered in confusion and misunderstanding," and that these confusions were "only partly due to the biases and passions engendered by race prejudices and consequent defense reactions." Dobzhansky believed that ultimately "the problem that now faces the science of man is how to devise better methods for further observations that will give more meaningful results."[21] Dobzhansky cited several approaches to the race concept that identify anywhere between five and thirty-two races, including a classification scheme proposed by Coon himself in the early 1950s that divided humanity into thirty races.

Dobzhansky was not critical of these inconsistencies in racial measurement. In fact, he embraced them because he believed that race itself was a tool for classification and sytematization, a device "used to make diversity intelligible and manageable."[22] While human genetic diversity is a hallmark of our species, Dobzhansky believed the way we choose to organize that diversity is a methodological decision and not one that necessarily reflects underlying evolutionary or hierarchical schemes. That is what race meant to Dobzhansky, and that is what he believed race meant in evolutionary biology and genetics.

That *Mankind Evolving* was published the same year as Coon's *Origins* attracted little notice at the time, in spite of the fact that the two books were basically about the same thing and reached two very different conclusions. This becomes even more puzzling when reading the chapters in *Mankind Evolving* that address, albeit indirectly, the theses of Coon's work. "We do not know nearly enough to give a satisfactory account of the appearance of modern man, *Homo sapiens*," wrote Dobzhansky, who then goes on to pick apart the separate creations argument.[23] First, Dobzhansky argues that autogenesis, the idea that human ancestors were "predestined from the beginning to develop in the direction of humanity" is not a feature of evolution.[24] Therefore, it is implausible that reproductively isolated groups of *erectus* would have evolved separately into *sapiens* over the course of 150,000 years. Second, because we recognize *Homo sapiens* as a single species, it could "not have arisen by the coalescence of two or several ancestral species."[25] Finally, Dobzhansky recognized that "living species are reproductively isolated, and living races are not" (humans cannot, for example, breed with chimpanzees or gorillas). This contradicted the view that Coon's five *erectus* populations around the world were isolated from one another and evolved separately.[26]

Coon, from his influential perch as the president of the American Association of Physical Anthropologists and as a well-respected leader in his field, challenged Dobzhansky's construction of race, and in doing so revealed its inherent contradictions. That the biological race concept could not completely disentangle itself from typological inferences was manifest in Coon's conflation of typology, genotype, and phenotype, as well as his own obvious social prejudices about nonwhites. In his review of *Origins*, which appeared alongside the Dobzhansky review in *Current*

*Anthropology*, Ashley Montagu acknowledged that race and typology were indeed inseparable, and that the race concept remained burdened by its connection to its own history. "If I were not actually reading the printed words in a book published in 1962 by a reputable publisher," Montagu wrote of *Origins*, "I could believe that I were reading an anthropology written and published in 1862."[27]

Just a year before the Coon imbroglio, Dobzhansky sparred with another anthropologist in the pages of *Current Anthropology*. Again, as he had been doing in both private correspondence and published writings, Dobzhansky defended the use of the race concept in the context of the evolutionary synthesis. In an article titled "On the Non-Existence of Human Races," University of Michigan anthropologist Frank Livingstone reasoned "that there are excellent arguments for abandoning the concept of race with reference to the living populations of *Homo sapiens*." For Livingstone, as one studied more and more of the variation both within and between populations, it was "nearly impossible to determine what the 'actual races really are.'" "To apply the term subspecies to any part of such variation not only is arbitrary or impossible but tends to obscure the explanation of this variation," Livingstone said. Ultimately, Livingstone's critique echoed the earlier arguments by Du Bois and Montagu. "No science can divorce its concepts, definitions, and theories so completely from its subject matter," he concluded.[28]

As a physical anthropologist, Livingstone had studied genetic diversity within and between human populations. His groundbreaking work on the sickle-cell trait in West Africa,[29] in fact, led him to reject racial classifications, as outlined in his back-and-forth with Dobzhansky. Dobzhansky, ever still the arbiter of the race concept, pushed back with his usual defense, arguing that "the difficulties with the race concept arise chiefly from failure to realize that while race differences are objectively ascertainable biological phenomena, race is also a category of biological classification." In other words, as Dobzhansky had argued many times before, race was real; it was the method of identifying and naming it where the problem lay. "There is nothing arbitrary about whether race differences do or do not exist, but whether races should or should not be named," Dobzhansky asserted, "and if they should, how many should be recognized, is a matter of convenience and hence of judgment."[30]

Though Livingstone agreed with Dobzhansky on this point, he ultimately thought this line of thinking—which had come to define the concept of race in biological terms—was specious. "I would agree with him [Dobzhansky]," Livingstone wrote, "that just about all human populations differ in the frequency of some gene" and that "all human populations are racially distinct, which should imply that each is a separate race." But in Livingstone's analysis, this made the term "race" "inapplicable" in a biological context. "Since any grouping differs with the gene frequency used," Livingstone argued, "the number of groupings is equal to the number of genes." Therefore, "one population could belong to several different races." This was, of course, Dobzhansky's point about the race concept: that it was a methodological tool to organize species-level diversity. But for Livingstone, this methodological justification for the use of race was not reconcilable with the taxonomic system—the Linnaean system—through which the biological sciences organized the natural world: "Any particular animal population cannot belong to several different genera, species, or any other taxonomic level within the Linnaean system; such a usage is inconsistent with the assumptions of the system."[31]

Later that decade, another dispute surfaced about the use of the biological race concept in the context of race, genetics, and intelligence—one that again highlighted the contradictions in the race concept. The 1969 publication of educational psychologist Arthur Jensen's "How Much Can We Boost IQ and Scholastic Achievement?" in the *Harvard Educational Review* became a lightning rod for scientific and popular discussion about race, intelligence, and genetics. In that article Jensen made his now-infamous claim that intelligence, or IQ, had a very high heritability, that that heritability was largely genetic in nature, and that therefore redress for discrepancies in IQ through educational remedies in order to "narrow the achievement gap between 'minority' and 'majority' pupils" was likely to be futile.[32] Jensen proposed instead that for American schools the "next step will be to develop the techniques by which school learning can be most effectively achieved in accordance with different patterns of ability."[33] In other words, trying to educate genetically smart students with genetically less-smart students is neither a good use of society's time or of humankind's genetic heritage. Jensen was hopeful that this was changing and believed that "we are beginning to see some definite signs that this mistreatment of the genetic basis of intelligence by social scientists may be on the wane, and that a biosocial view of intellectual development more in accord with the evidence is gaining greater recognition."[34]

In the article, Jensen, an educational psychologist with no formal training in any area of the biological sciences, presented himself as an expert in genetics and tried to summarize for the psychological community the mechanisms of heredity and how they related to race and IQ. Sections of the paper titled "The Concept of Heritability," "Polygenic Inheritance," "Covariance of Heredity and Environment," "Common Misconceptions About Heritability," and "Empirical Findings on the Heritability of Intelligence" help acquaint the reader with complex genetic concepts. Attempts like Jensen's to link genetics with race and America's social conditions were part of the academic and popular culture in the 1960s. This was occurring just as the civil rights movement and the more radical Black Power movement were quickly moving beyond a strictly civil rights platform to pay close attention to broader issues of social and economic justice.[35]

In 1966 William Shockley, a Nobel Prize–winning physicist and professor at Stanford University, received considerable publicity for calling upon the National Academy of Sciences (of which he was a member) to "foster research on the effects of heredity, including race, on human behavior." "I evaluate the marrow of the city's slum problem to be our uncertainty about its genetic aspects and our fear to admit ignorance and to search openly for relevant facts," Shockley claimed to the academy in a paper with the title "Possible Metallurgical and Astronomical Approaches to the Problem of Environment versus Ethnic Heredity."[36] In a press release by the Stanford University News Service, Shockley explained his "incursion into the field of genetics as a reaction to colleagues' opinions that 'even the proposal that an objective study be sought is doomed to be smothered under an emotional slogan.'"[37] The following year, at a speech at Michigan State University, Shockley warned that "unrealistic hope for speeding the Negro's struggle for equality before the real causes of his disadvantages are known may unwittingly inflict untold human suffering on the Negroes themselves."[38] With his proposed program going nowhere at the National Academy, Shockley called upon John Gardner, Lyndon Johnson's secretary of health, education, and welfare, to undertake his study of America's impending racial degeneration. Gardner passed the letter off to Philip Lee, assistant secretary for health and scientific affairs, who rebuked Shockley's request and suggested that he speak with his Stanford colleague the eminent geneticist Joshua Lederberg about genetics, the environment, and human behavior.[39]

The attempt by men like Jensen and Shockley to employ a biological race concept shows that no matter how hard Dobzhansky and other biologists tried to narrowly define race in the context of biology—that despite the stated shift away from typology to populationist thinking—race could and would be used for typological, racist, and nonscientific ends. The Jensen and Shockley affairs confirm the plasticity of the biological race concept and that its historical legacy is inseparable from its contemporary usage. At a 1966 symposium held at the annual meeting of the American Association for the Advancement of Science exploring science and the race concept, the giants of evolutionary biology and genetics, plus several esteemed psychologists and anthropologists, continued to argue about the meanings of race in science. The symposium was organized, in the words of the anthropologist Margaret Mead, "as an indication of the present state of knowledge and research on problems of race, and it represents a response to the barrage of pseudoscientific statements which, since the Supreme Court desegregation decision of 1954, have attempted to prove the innate biological inferiority of the group of Americans who are socially classified as Negro."[40] At the symposium Dobzhansky, Mayr, and other evolutionary biologists again insisted that the biological race concept had left typology behind, and that race was both a tool for classification and a biological phenomenon. It is in that space, between methodology and biology, where the contradictions of race would emerge again and again.

One wonders, despite Dobzhansky's stern protests to Montagu that abandoning the term "race" in biology would do nothing—"I have done my damndest to convince you that this proposition will neither be accepted nor would it do any good if accepted"—what might have been had geneticists and others in the biological sciences heeded the call to substitute the term "ethnic groups" for "race."[41]

# 10

## NATURALIZING RACISM

---

### The Controversy Over Sociobiology

O n December 18, 1975, after a long illness, the biologist Theodosius Dobzhansky died. As an architect of the evolutionary synthesis in the biological sciences, a distinguished teacher and mentor to a generation of students at Caltech, Columbia, and Rockefeller, and a public intellectual, Dobzhansky left an indelible mark on his discipline, on the sciences more generally, and on the times in which he lived. Eulogized by the Royal Society, Dobzhansky was remembered for his singular contributions to the field and as "a man ready to adjust his views in the light of increasing knowledge and not ashamed to admit past error, nor concerned to disguise it."[1] What Darwin's *The Origin of Species* was to evolutionary theory in the nineteenth century Dobzhansky's *Genetics and the Origin of Species* was to it in the twentieth century according to evolutionary biologist and former Dobzhansky student Francisco Ayala.[2] Another student of Dobzhansky's, the evolutionary biologist Richard Lewontin, remembered his late mentor as the scientist "who, in *Genetics and the Origin of Species*, changed our view of races from the taxonomist's typological classification to the modern population genetic one."[3] Indeed, Dobzhansky's impact is historic.

Just a year before his death, Dobzhansky had published a remembrance of his close friend and longtime collaborator L. C. Dunn, who had died in March of 1974. Like Dobzhansky, Dunn was a giant in his field, remembered both for his important contributions to genetics and for the steadfast humanism that he brought to his work in science and society. Dobzhansky noted that during the 1930s Dunn was active on the Emergency Committee for German Scholars that helped resettle German exiles at American universities.[4] Dunn was also remembered as an authority on the race concept—work that he hoped would undercut racism's scourge in both scientific thought and popular practice.[5] Other members of their scientific cohort, also founders of and contributors to the evolutionary synthesis, died in the decade or so following Dunn and Dobzhansky. Julian Huxley passed in 1975. Curt Stern died in 1981, George G. Simpson in 1984, and Sewell Wright in 1988. There were, however, two long-lived exceptions. Ashley Montagu and Ernst Mayr remained prolific into their nineties; Montagu died at age ninety-four in 1999, and Mayr at age one hundred in 2005.

## SOCIOBIOLOGY ASCENDANT

It is purely coincidental that in the same year as Dobzhansky's passing the entomologist Edward Osborne Wilson published *Sociobiology: The New Synthesis*, a now classic work examining the relationship between biology and social behavior in all species, including humans.[6] Before his foray into sociobiology, Wilson had a distinguished career as a biologist whose work on ant populations, biogeography, and ecology made him one of the most distinguished and publicly recognized scientists of the late twentieth century. Wilson, of course, would also go on to become one of the world's most prominent environmental advocates as a voice for the preservation of the earth's biodiversity. But it was his work on sociobiology that first thrust him into the scientific and public spotlight. In the years following *Sociobiology*'s publication, Wilson would, for example, win a Pulitzer Prize for *On Human Nature* and a National Medal of Science from President Carter.

Wilson's sociobiological proposition—that a biological theory of behavior (in humans and nonhuman species) should inform the work of the social sciences and humanities (some called this disciplinary imperialism or the Darwinizing of the social sciences[7])—can now be understood as part of the impulse to unify the biological sciences.[8] The most important step in this struggle during the first half of the twentieth century was the emergence of the evolutionary synthesis—the unification of Darwinian natural selection with Mendelian genetics. In declaring the "new synthesis," Wilson sought to attach *Sociobiology* to this legacy, applying evolutionary concepts to the relationship between biology and behavior in all species. As Wilson wrote, his work sought to "codify sociobiology into a branch of evolutionary biology and particularly modern population biology."[9] In addition to its historical connection to the drive in biology to unify related disciplines, sociobiology's ambition to uncover "the biological basis of all forms of social behavior in all kinds of organisms, including man," was heir to a legacy within the biological sciences of integrating claims about the genetic basis for complex human social behavior into social thought.[10]

While sociobiology made no claims to immediately involve itself in social policy, Wilson's introductory chapter in *Sociobiology* did propose that "the principal goal of a general theory of sociobiology should be an ability to predict features of social organization from a knowledge of these population parameters combined with information on the behavioral constraints imposed by the genetic constitution of the species."[11] A leap from predicting the features of social organization to influencing it was not beyond the ambitions of some sociobiologists, including Wilson. If "the most diagnostic features of human behavior evolved by natural selection and are today constrained throughout the species by particular sets of genes," then sociobiology's contribution to human society would someday be, according to Wilson, to "monitor the genetic basis of social behavior" because "the genetic foundation on which any such normative system is built can be expected to shift continuously."[12]

The quest for a sociobiologicalization of the biological sciences was not new to academic or popular thinking. Darwin's *The Descent of Man*, and later R. A. Fisher's *The Genetical Theory of Natural Selection* and Julian Huxley's *Evolution: The Modern Synthesis*, expressed varying levels of belief in the biological basis for human social behavior.[13] And, to be sure, Wilson was neither the first nor the only proponent of sociobiology in the late twentieth century. Sociobiology with a small *s* was a field that grew from the achievements of scientists working in a range of biological and behavioral disciplines. For example, the ethological work of Konrad Lorenz and Desmond Morris in the 1960s laid the foundation for later sociobiological research. In the 1970s, Richard Dawkins, author of the best-selling *The Selfish Gene*, and David Barash, author of *The Whisperings*

*Within*, were among the many natural and social scientists working from a sociobiological paradigm.[14]

In the five years following the publication of Wilson's tome, both the academic and popular presses closely followed sociobiology and the debate over its scientific merits and political significance. Through an aggressive public relations campaign, general interest in sociobiology gained a societal salience resonating from the hallowed halls of academia to the glossy pages of *People* magazine. *Business Week* called the new science a "genetic defense of the free market," while *Psychology Today* asked whether Wilson's synthesis was "all wet?" declaring that it could "provide very little enlightenment about the behavior of real people."[15] University of Chicago cultural anthropologist Marshall Sahlins agreed, attacking the sociobiological notion that human kinship relations are "organized in accord with the genetic coefficients of relationship as known to sociobiologists." Sahlins insisted, "Human social behavior is not organized by the individual maximization of genetic interest in that human beings are not socially defined by their organic qualities." Sahlins argued that humans were biological beings but that culture, not biology, "is the indispensable condition" of human organization.[16]

Others embraced *Sociobiology*. Writing in the *New York Times*, the anthropologist John Pfeiffer said the book "has much to say about what is happening to us here and now," and that "the book may be regarded as an evolutionary event in itself, announcing for all who can hear that we are on the verge of breakthroughs in the effort to understand our place in the scheme of things."[17] In the *Quarterly Review of Biology*, the biologist Mary Jane West-Eberhard called the book "brilliant and timely."[18] David Barash deemed it a "tour de force" and favorably compared Wilson and his work to Julian Huxley's *Evolution: The Modern Synthesis*.[19] In private correspondence several distinguished scientists echoed these sentiments. Konrad Lorenz wrote to Wilson that he had read the book "with as great pleasure as gain." Philip Darlington, the Harvard biologist who had trained Wilson, called *Sociobiology* a "fundamental, magnificent, and useful piece of work."[20]

Most of *Sociobiology*, according to Stephen Jay Gould, both a critic of the field and a friend of Wilson's, "wins from me the same high praise almost universally accorded to it." Gould, like many other champions of the field, believed Wilson's work to be "a lucid account of evolutionary principles and an indefatigably thorough discussion of social behavior in all groups of animals," and predicted that *Sociobiology* would be "the primary document" on these matters "for years to come." However, it was Wilson's (and, of course, others' in the field) application of sociobiological ideas to humans that came under fire, particularly for the way in which these ideas were presented. It was what Gould called Wilson's "extended speculation on the existence of genes for specific and variable traits in human behavior—including spite, aggression, xenophobia, conformity, homosexuality, and the characteristic behavioral differences between men and women in Western society" that became a target for sociobiology's critics.[21] It is the field's interest in what it deemed human universals, like xenophobia and ethnocentrism, that is particularly noteworthy in the context of the evolution of the race concept and its implications for understanding racism.

## SOCIOBIOLOGY AND RACE

Dobzhansky's *Genetics and the Origin of Species* modernized a theory of *race* in the context of the evolutionary synthesis and population genetics. In a similar way, Wilson's *Sociobiology* and contemporary sociobiological works sought to articulate a modern theory of the existence of *racism* in the context of the new synthesis. Wilson was no racist and was distressed by the uses of sociobiology in this way, but in making arguments for an evolutionary and genetic basis for xenophobia and ethnocentrism, sociobiologists were participating in a long-standing debate regarding biological explanations for these concepts. Although the new synthesis did not contest the race concept as described by the modern synthesis, it did offer a biological way to explain tensions that persisted between racial (and other) groups in American society that resonated in the wake of the social revolutions of the 1960s. Literary critic Henry Louis Gates has called this sociobiology's "pessimism," that "racism is in our nature" and that "little can be done about it."[22]

The 1960s and early 1970s were part of an era in the United States in which racial movements for social justice radically pushed back against and began to change America's political and cultural arrangements.[23] As sociologists Howard Winant and Michael Omi point out, the wake of the 1960s social movements "constituted a period of profound social transformation and dislocation in American life."[24] In the 1970s, that transformation and dislocation were manifested in a renewed backlash against the civil rights struggle. Sociobiology gained social salience at this historical moment, and the biological explanations for xenophobia and ethnocentrism, as articulated by some in the field, were well suited to a post–civil rights movement age witnessing a rise in cultural nationalism and ethnic pride and a growing rejection of the integrationist ideal.[25]

This was also an era that saw a dramatic rise in genetic explanations for complex human social behaviors. Sociologists Dorothy Nelkin and M. Susan Lindee, in *The DNA Mystique: The Gene as a Cultural Icon*, argue that since the late 1970s "in many popular narratives, individual characteristics and the social order both seem to be direct transcriptions of a powerful, magical, and even sacred entity, DNA." Genes, they say, are a "convenient way to address troubling social issues: the threats implied by the changing roles of women, the perceived decline of the family, the problems of crime, the changes in ethnic and racial structure of American society, and the failure of social welfare programs."[26] As sociologist Troy Duster shows, "both the popular media and scientific journals published an explosion of articles that staked a renewed claim to the genetic explanation of matters that the previous two decades had laid to rest as social and environmental." For example, according to Duster, a review of the *Reader's Guide to Periodical Literature* from 1976 to 1982 shows a surge in articles that "attempted to explain the genetic basis for crime, mental illness, intelligence, and alcoholism during this brief six-year period."[27] Sociobiology played an important role in the popularization of these ideas.

Wilson's toughest critics acknowledged that sociobiology was not immediately concerned with race, nor was it itself a racist doctrine. However, even if sociobiology was definitively not a racist theory, in its pursuit of a biologically based theory of human social behavior it *did* concern itself with race indirectly. By trying to describe why, in an evolutionary and genetic way, populations of peoples hated, feared, and distrusted one another, sociobiology contributed to a biological concept of racism. So while the biological race concept was largely ignored in sociobiological theory, the social meanings of race became a focus of sociobiological research through the study of the biological mechanisms of racism. In this context, sociobiology could be and was interpreted by some as a way to rationalize racism. Racists, of course, did not need scientific theories like sociobiology to promote their ideas. Nevertheless, the intellectual claims of sociobiology—intentionally or not—could and did serve the needs of those who harbored racist ideas by giving them scientific legitimacy.

Because Wilson was a distinguished biologist trained in the traditions of the evolutionary synthesis and population biology, *Sociobiology* did nothing to challenge the race concept: its definition had largely been settled in this corner of biology by the time the book was written. Wilson even turned to his Harvard colleague and soon-to-be antisociobiology antagonist Richard Lewontin for authority in this area, citing his work on the race concept in biology and writing in *Sociobiology* that "human populations are not very different from one another genetically."[28] Furthermore, because *Sociobiology* sought to identify human universals, it was not, in itself, a racist doctrine—a fact that even Wilson's most vociferous critics acknowledged. After all, as Lewontin acknowledged, Wilson was concerned with the "universals of human nature—his chief point is that we're all alike."[29] This was a point many sociobiologists made as well. The British sociobiologist and anthropologist Vernon Reynolds pointed out that "sociobiology expects *every* racial group to be xenophobic and it makes *no* statement at all about superiority."[30]

Unlike sociobiology, earlier biologically driven theories of human nature did have race at their core. At the end of the nineteenth century, for example, social Darwinists spoke of the inability to "change the different qualities of the races." In the early twentieth century eugenicists spoke of "feebleminded" blacks and other groups, while IQ theorists of the 1960s and 1970s speculated about the relationship of race, genetics, and IQ.[31] Sociobiologists, however, consciously sought to sever any connection between their discipline and any racist doctrine. In his Pulitzer Prize–winning work *On Human Nature*, for example, Wilson issued a "strong caveat" declaring the search for "racial differences in behavior" as "the most emotionally explosive and politically dangerous of all subjects."[32]

For a more robust understanding of a sociobiological discussion of racism, we must return briefly to the basic theoretical suppositions of sociobiology. Sociobiologists believe that "animals are social to the extent that cooperation is mutually beneficial." Sociality exists, therefore, because of kin selection—"the selection of genes due to one or more individuals favoring or disfavoring the survival and reproduction of relatives (other than offspring) who possess the same genes by common descent." In turn, cooperative social behavior among close kin will maximize inclusive fitness, or "the total effect of kin selection with reference to an individual." Some sociobiologists assert that both ethnicity and race are "extensions of the idiom of kinship, and that, therefore, ethnic and race sentiments are to be understood as an extended and attenuated form of kin selection."[33] Because kin groups are both aggressive and xenophobic, according to sociobiological theory, one can assert, as some sociobiologists do, that xenophobia, ethnocentrism, and racism have a genetic basis.[34]

In Wilson's seminal text he articulates a sociobiological explanation for how humans make sense of difference. Modern humans inherited a xenophobic trait from our hunter-gatherer ancestors, according to *Sociobiology*, and "human behavior provides some of the best exemplification of" what Wilson calls "the xenophobia principle": a "newcomer is a threat to the status of every animal in the group, and he is treated accordingly," and "cooperative behavior reaches a peak among the insiders when repelling such an intruder." Offering a strikingly simplistic view of intergroup hatred, Wilson argues that in humans "outsiders are almost always a source of tension." When they "pose a physical threat . . . they loom in our vision as an evil, monolithic force." Furthermore, Wilson speculates that the outsiders are reduced "to subhuman status, so that they can be treated without conscience. They are the gooks, the wogs, the krauts, the commies—not like us, another subspecies surely, a force remorselessly dedicated to our destruction who must be met with equal ruthlessness if we are to survive." Wilson hypothesizes that this occurs at the level of a "gut feeling," and he compares the xenophobic trait of a human being and that of a rhesus monkey, suggesting that the behavior "may well be neurophysiologically homologous."[35]

Pierre van den Berghe, trained as a sociologist, was one of the most prolific proponents of the sociobiology of ethnocentrism and xenophobia. A specialist in race and ethnic relations and a professor at the University of Washington, van den Berghe had asserted in his presociobiologically influenced work that "racism is ultimately reducible to a set of attitudes which are, of course, socially derived but which nevertheless become part of the personality system of the individual."[36] However, in the mid-1970s van den Berghe's approach shifted, and there emerged in his work a belief in the sociobiological origins of race relations. Van den Berghe suggested, for example, that "ethnocentrism evolved during millions, or at least hundreds of thousands of years as an extension of kin selection."[37]

Van den Berghe and Wilson were not alone in their sociobiological explanations for human intergroup antagonism. Throughout the sociobiology literature ethnocentrism and xenophobia are justified in sociobiological terms. Another of the early sociobiologists, the zoologist David Barash (also of the University of Washington), posits a theory similar to van den Berghe's, but with a twist. Barash believes that "human beings have a notable inclination to exclude individuals who are conspicuously different from us in any way." The root of this behavior is biologically adaptive "for the following reason: Among most animals, disease is a prominent cause of mortality. . . . Since many diseases can be transmitted by infected individuals, it would be advantageous if diseased animals were somehow prevented from associating closely with the healthy ones. Thus, in many animal societies, diseased or disfigured individuals are often mercilessly hounded and excluded from the group. As a general rule, those that are different get ostracized."[38] Barash offers little hope of "evolving" out of this predicament. Instead, as he asserts, we must "make sure that we are carefully taught to love one another." Sociobiology is thus "an antidote to racism" because "it emphasizes human universals" (e.g., human nature) and thereby unites the species in "biological oneness."[39]

Not all scientists who wrapped themselves in the mantra of sociobiology were as kind. Psychologist J. Philippe Rushton was by far the most extreme of the sociobiologists, advocating not only sociobiological notions of racism but also typological notions of race. One might argue that his extreme views have no place in sociobiology. Rushton's work, however, has appeared in sociobiological and other respected scientific journals.[40] He even proclaims his work sociobiological in nature in his monograph *Race, Evolution, and Behavior*. Furthermore, he was the recipient of several distinguished fellowships, including a Guggenheim, which only served to increase his legitimacy. Rushton's work, then, perhaps can be seen as a bridge between eugenics and the mainstream realm of sociobiology.

Rushton's ideas were echoes of the works of Gobineau and Morton. He was convinced that cranial capacity, which he found to be smallest among blacks, had direct correlations with "intelligence, social organization, sexual restraint, and quiescent temperament." He ascribed the origins of these differences to the Ice Age, which exerted selective pressures upon non-"Negroid" races. Rushton's work also attempted to rekindle the debunked myth of r/K selection, where r is high reproduction rates and K is high levels of parental investment. According to Rushton's conclusions, r-selected reproductive strategies are emphasized among blacks, thus causing a "bioenergetic trade-off between which is postulated to underlie cross-species differences in brain size, speed of maturation, reproductive effort, and longevity." K-selected races have higher intelligence and thereby more advanced cultures. This notion of selective pressure as applied to human social behavior is similar to sociobiological notions of gene-culture coevolution. It is interesting that Rushton's work is cited numerous times in Richard Herrnstein and Charles Murray's The Bell Curve: Intelligence and Class Structure in American Life, the mid-1990s best seller that linked race, genetics, and intelligence.[41] Herrnstein and Murray nod approvingly toward Rushton's work, writing that "he has strengthened the case for consistently ordered race differences." Nonetheless, they do acknowledge that his "theory remains a long way from confirmation."[42]

Rushton, to be sure, is far more extreme in his application of coevolutionary ideas than most sociobiologists. His work is also tarnished by his association with the Pioneer Fund—an openly racist organization that supported his research and that he would go on to head from 2002 to 2012. The fund—founded by Wickliffe Draper, whose role in the eugenics movement and in scientific racism is documented in chapter 5 of this book—is a eugenic foundation that provides approximately $1 million a year to academic researchers. Much of the work from its grants goes toward establishing the genetic bases for racial differences.[43]

There were sociobiologists who were skeptical of the extremes of Rushton. Writing in 1986, Vernon Reynolds, Vincent Falger, and Ian

Vine cautioned, "Some attempts to apply the sociobiological paradigm to human behavior have certainly been premature, ill-considered, superficial, over-confident, and indeed socio-politically naïve." They were particularly concerned that sociobiological work could mistakenly portray racism as one of the "genetic imperatives of human nature." Others, like Robin I. M. Dunbar (despite his support for studying the sociobiology of ethnocentrism), were concerned about the scope of sociobiological claims given "the difficulty with producing evidence in support of any sociobiological statement with long-lived species like our own."[44]

## THE SOCIOBIOLOGY DEBATES

In the months and years that followed the publication of *Sociobiology* and the rapid growth of the field itself, the debate over the field grew increasingly personal and was sometimes outright nasty. Wilson, as the field's most distinguished member, quickly became both its spokesperson and lightning rod. Tense correspondence between Wilson and his critics reveals deep scientific and political acrimony on both sides of the debate. Some of the bitterest animosity emerged between Wilson and several of his colleagues at Harvard, particularly Richard Lewontin. Wilson claimed the attacks against him were personal in nature (as they sometimes were), and he even recused himself from faculty meetings in the Department of Organismic and Evolutionary Biology at Harvard. He found "it intolerable to sit at a department-level meeting where the chairman [in reference to Lewontin] is the same individual who is conducting a personal campaign against me and my work, largely on political grounds."[45] Lewontin, in fact, had been very aggressive and unsparing in his criticism of *Sociobiology*. "The book has a lot of science *in it*, but it is not *of* science," Lewontin wrote to Wilson in October of 1975 in one of the many heated letters they exchanged following *Sociobiology*'s publication. Concerning the scientific status of *Sociobiology*, Lewontin told Wilson that his book was "not a work of science because it makes use of what is essentially an advocacy method to erect an untestable and non-falsifiable structure of explanation of a world that in many respects does not even exist."[46]

Wilson expressed his disbelief in the initial reaction to his work, concluding that most of the attacks on his and other sociobiological works were partisan, and he attempted to offset the impact of his critics by dismissing them as ideological dogmatists. While Wilson did, however, recognize what he called his "more moderate" critics, he criticized them too for holding fast to the "antigenetic bias that has prevailed as virtual dogma since the fall of Social Darwinism."[47] Accused of creating a scientific theory that sought to buttress the status quo, Wilson defended himself as a liberal who had "no interest in ideology." His "purpose was to celebrate diversity and to demonstrate the intellectual power of evolutionary biology."[48]

Wilson's critics saw his work quite differently. In 1975, a letter to the editor in the *New York Review of Books* authored by members of the Sociobiology Study Group of Science for the People (a political group that included several of Wilson's Harvard colleagues) attacked *Sociobiology* and Wilson, calling the work "a particular theory about human nature, which has no scientific support, and which upholds the concept of a world with social arrangements remarkably similar to the world which E. O. Wilson inhabits." The letter associated Wilson's work with earlier biologically deterministic theories, including social Darwinism and eugenics, stating that such theories (specifically *Sociobiology*) "provide a genetic justification of the status quo and of existing privileges for certain groups according to class, race or sex."[49] Wilson wrote to the *New York Review of Books* editor Robert Silvers, claiming that the article represented "the kind of self-righteous vigilantism which not only produces falsehood but also unjustly hurts individuals and through a kind of intimidation diminishes the spirit of free inquiry and discussion crucial to the health of the intellectual community."[50] Deeply hurt by what he believed were more than just scientific or partisan attacks against *Sociobiology*, Wilson's anger at what he perceived to be indecorous and inappropriate behavior for the academy certainly drove his intense response to his critics. Both stunned by and incredulous at the nature of the attack against him and his book, Wilson wrote to Lewontin that December (1975) asking, "How can you make such repeated accusations which you *know* are untrue, especially since the target is an old and close associate?"[51]

Not all Wilson's collegial relations at Harvard were distressed by the sociobiology debates. In November 1975, Wilson had written to the Harvard paleontologist and evolutionary biologist Stephen Jay Gould—also a critic, but one with whom, unlike with Lewontin, Wilson kept up a good personal relationship throughout the rancor that followed the book's publication. In his note, Wilson enclosed his response to the *New York Review* article and hoped that Gould would agree that had he shown him "this elementary courtesy, your own letter would have been far different and my reputation would not have been unjustly damaged." Wilson also told Gould that he knew that what he called the incident was ideological in origin and that there was "no personal animosity between us."[52] Throughout the sociobiology debates, Gould and Wilson maintained good relations. Gould even defended Wilson against some of the more personal attacks later directed against him.[53]

Wilson defended his work in a 1976 article titled "Academic Vigilantism and the Political Significance of Sociobiology." He believed that *Sociobiology* was being judged "according to whether it conforms to the political convictions of the judges, who are self-appointed" and that those found guilty by such judges, like himself, were "given a label and to be associated with past deeds that all decent persons will find repellent." Wilson defended his foray into human sociobiology, claiming that it "was approached tentatively and in a taxonomic rather than political spirit" and that it "was intended to be a beginning rather than a conclusion."[54] Nonetheless, Wilson clearly underestimated the political and historical context into which he offered these theories, and how that context would shape the tone and intensity of the response to both his work and to sociobiology in general.

Among the first to react to the publication of *Sociobiology* was the Sociobiology Study Group of Science for the People, which criticized sociobiology as a "tool for social oppression." The study group's purpose was to "publicize the view that *Sociobiology* was just another form of biological determinism and to break through the solid mass of praise of the book and make the subject controversial" as well as to produce "ideological weapons to counteract and delegitimate sociobiology and biological determinism in general."[55] Science for the People, a coalition made up largely of socialist and ideological Marxist biologists, scholars, and community activists, worked "to analyze and combat" what it called "this latest appearance of biological determinism." The group's stated goals were (1) "exposing the class control of science and technology; (2) organizing campaigns which criticized, challenged, and proposed alternatives to the present uses of science and technology; and (3) developing a political strategy by which people in the technical strata could ally with other progressive forces in society."[56] Among the study group's members were scientists at Harvard and other Boston-area universities, including Lewontin, Gould, and biologist Ruth Hubbard.[57]

The group's reaction to the publication of *Sociobiology* and other sociobiological theories was often swift and unsparing of the field and its nominal leader. Its attacks on sociobiology appeared in the group's bimonthly publication *Science for the People*, as well as in numerous journals, newspapers, and periodicals. When sociobiologists proclaimed themselves reductionist and deterministic, asserting that a biologically based human nature was inherent to the human genome, the study group responded by stating, "Determinism provides a direct justification for the status quo as 'natural.'"[58] When sociobiologists asserted that "even with identical education and equal access to all professions, men are likely to play a disproportionate role in political life, business, and science," the group argued that evidence for male domination as a facet of human nature was wrongheaded, ignoring or misrepresenting "cross cultural evidence on sex roles" that dispute sociobiology's ideas.[59] Finally, when sociobiologists claimed compelling genetic evidence for their assertion that there exist human universals such as spite, aggression, male dominance, ethnocentrism, and territoriality, the study group maintained that sociobiologists had no evidence for their theory. "It is the classic error of reification," said members of the group, "by which mental constructs are treated as if they were something concrete or object."[60]

The group struggled, however, with whether *Sociobiology* was racist. Minutes from the group's June 1977 meeting suggested a divide among its members: "Some felt the book is clearly racist, others that it is not really a serious issue in the book, others that there is racism but it is implicit." A discussion on the subject with the radical group Committee Against Racism (CAR) yielded little agreement, with CAR convinced that sociobiology was racist, arguing that "ideologies which can be used to justify racism are racist." Another group member expressed the belief that "since *Sociobiology* helps to legitimize racial conflict by postulating that it is genetically based, it helps to promote racial conflict and do what racism does indirectly."[61]

If Wilson was the public face of sociobiological theory, then Lewontin had quickly become the public face of opposition to that theory. Even though he was a self-proclaimed ideological Marxist, he was still, like Wilson, one of the nation's preeminent biologists. But unlike Wilson, Lewontin explicitly fused his scientific activities with his political ones. And despite his protests that he should not be considered the voice of Science for the People, he became just that, precisely because of his intellectual standing in the scientific community and his long history of engaging in what he believed were the misuses of science for political and social purposes. Lewontin claimed that "no one is a spokesman for SftP," and that calling him such misunderstood the organization, which he described as "nonhierarchic and virtually anarchistic [in] nature."[62]

Members of the Sociobiology Study Group were not the only ones to oppose sociobiology, and although Wilson might have accused them of "academic vigilantism," their tactics in opposition to sociobiology paled in comparison with more radical groups that had threatened and committed acts of violence against Wilson. At the 1978 American Association for the Advancement of Science's annual meeting, for example, the International Committee Against Racism (INCAR), a radical Marxist organization, charged the stage as Wilson was about to speak. The protestors doused him with ice water and then, with antisociobiology placards in hand, shouted, "Racist Wilson you can't hide, we charge you with genocide!!!"[63]

Apart from the attacks by INCAR in public settings, most critics of Wilson's engaged him intellectually on the scientific content in *Sociobiology* and on its relevance as a political document, even as those criticisms devolved into personal attacks on both sides. Yet over time Wilson continued to engage his academic critics, particularly his Harvard colleague Lewontin, in much the same way he reacted to his purely political ones: he attacked their political proclivities rather than engaging them in scientific debate. He even accused his critics of trying to thwart academic freedom.[64] And this strategy was largely successful—Wilson succeeded in deflecting scientific criticism of his sociobiological work by accusing many of his critics of being Marxists (which, indeed, they were). In reaction to a published interview of Lewontin in the French newspaper *Le Monde*, Wilson wrote to the editor referring to Lewontin as "my chief Marxist critic."[65] Wilson also suggested that opposition to sociobiology among his Harvard colleagues and Science for the People critics was driven by what he believed was their fear that Marxism was "mortally threatened by the discoveries of human sociobiology."[66] Overall, the conflict between Lewontin and Wilson, as well as the debates over sociobiology, were part of a larger cultural trend in academia in the 1960s and 1970s in which liberals and Marxists clashed over what values should shape their work and their world.[67] This dispute is also, of course, a glaring example of how political considerations can shape scientific researchers and the questions they pursue in their work.

Wilson's use of Marxism as an intellectual foil to *Sociobiology* holds great irony for reasons best articulated by Wilson himself. "Marxism is sociobiology without biology," Wilson wrote in 1978, just a few years after *Sociobiology*'s publication. Whereas Marxists view the world through a lens of class struggle, biological and genetic mechanisms were critical to the sociobiological worldview. Wilson's popularized accounts of sociobiology and his public statements on the nascent science often began not with a methodological defense of his data or theoretical claims but rather with an outline of the controversy surrounding sociobiology.[68] Wilson spent, for example, an entire chapter of his 1983 book *Promethean Fire: Reflections in the Origin of Mind*, attacking critics of sociobiology and accusing Lewontin and other critics of promoting "Marxism-Leninism, in a manner specialized to subordinate science to the service of that [Marxist] ideology."[69]

Even accepting Wilson's assertion that he was a political neophyte at the time of *Sociobiology*'s release, he quickly became an apt pupil of his discipline's politics, of academic politics more generally, and certainly of society-wide politics. Wilson, for example, kept a file titled "Political Uses of Theory by New Right." Richard Lynn, a noted racist and eugenicist who was then teaching at the New University of Ulster in Great Britain, wrote to Wilson that "several of us in Ulster have been greatly impressed by your book *Sociobiology* and also sorry to see you coming under attack from 'liberals.' Do not be discouraged. . . . Many of us think sociobiology reasonable and sensible." Enclosed in that correspondence was an article by Lynn called "The Sociobiology of Nationalism," published in July 1976. In it Lynn concluded, "Separatism promotes genetic variety because it divides a population into a number of smaller, inbreeding populations. Each of these develops its own genetic characteristics."[70] Similarly, in 1977 Wilson received a letter from a Wilmot Robertson, who enclosed an article from *Instauration Magazine* (a white supremacist publication) with the title "The Minority War on Science." "You may think we come on a little too strong," Robertson wrote, "but I do feel there is a place in the United States and in the world for one magazine that looks at the racial background of the battles now going on in the scientific profession." "If all races had an equal capability for science," the *Insaturation* piece read, "there should be no marked differences in scientific methodology when one replaced another in a dominant position in science. If it can be shown, for example, that Western science is changing its habits and its practices as Jews achieve an ever more commanding position, then this ought to demonstrate that science in its essentials is not universal, not a body of ideas and methods that stand above nation and race, but on the other hand is dependent on race for its very character and its very existence."[71]

Lewontin also drew Wilson's attention to the relationship that was developing between sociobiological and racist theory. "I thought you might be interested in the enclosed," Lewontin wrote to Wilson in July 1979, "which comes from the Journal of the National Front in Britain." The article, by Richard Verrall, whom Lewontin mockingly referred to as the National Front's "resident 'intellectual,'" was titled "Sociobiology: The Instincts in Our Genes." Lewontin quotes the article, which describes the sociobiological nature of racism: "The basic instinct common to all species to identify only with one's like group; to inbreed and to shun out-breeding. In human society this instinct is racial, and it—above all else—operates to ensure genetic survival. The modern races of man evolved in pre-historic times, the separate development of each race representing an evolutionary experiment which nature isolated by instinctive tribal antagonisms."[72]

Not to miss an opportunity to point out to Wilson how Wilson's anti-Marxist jabs had angered him, Lewontin wrote, "Your red-baiting makes you a natural ally of this type, but then I guess I heard somewhere about how if one lies down with dogs, one gets up with fleas."[73] The Sociobiology Study Group had also debated how the group would respond to the National Front's racist assault. One member was concerned that a "description of sociobiology as racist" would only encourage the National Front "to write more on sociobiology." The group also wondered if its statement would "force Wilson to make a disclaimer in a British journal."[74] Wilson was not pleased that his sociobiological work was being used in this way. In a handwritten note found in his personal papers in a file labeled "Political Uses of Theory by New Right," Wilson wrote, "Radical right literature 1981. Worse than radical left!"[75]

For his part, Lewontin was well practiced at being a scientific contrarian. Just a few years before he mounted a challenge to sociobiological theory, Lewontin withdrew his membership in the prestigious National Academy of Sciences in protest of what he believed was the academy's inappropriate involvement in war-related research.[76] In addition to his contribution to the sociobiology debates, Lewontin was involved in the debates over the relationship between intelligence and race in the early 1970s. Some of Lewontin's arguments against those advancing a connection between heredity, IQ, and race anticipated his rhetoric against Wilson and others in the sociobiology debates. The Harvard psychologist Richard Herrnstein, following the 1973 publication of his book *IQ and the Meritocracy*, threatened Lewontin with a lawsuit for his participation in an advertisement placed in both popular and academic media. The ad criticized Herrnstein's IQ research as racist, a characterization that he felt was "repulsive and slanderous," as well as a "gross misrepresentation of my writings." In response to Herrnstein's threatening letter, Lewontin wrote, "I do not want to defame you, but to expose the fallacy and pernicious effects of certain ideas that you have propagated." "I do not accuse you of being *primarily* a racist," Lewontin asserted, "but whatever your motivation, your activities *are* promoting the cause of racism."[77]

Also in the early 1970s, Lewontin received similar correspondence from the physicist William Shockley, who similarly demanded the retraction of comments Lewontin had made against him concerning his public statements and publications about African Americans and intelligence. Whereas Lewontin tried to muster as much respect as he could for Herrnstein in their correspondence, he was unable to hide his sarcasm and disdain in his exchange with Shockley. "I am in receipt of your communication demanding that I make unspecified changes in the remarks in the *Harvard Crimson*," Lewontin wrote. He goes on to tell Shockley that he was not sure what he had had in mind in terms of changes, but that after rereading what he had said in the article, Lewontin himself felt dissatisfaction at his comments and therefore would take this opportunity to send a new statement to the editor of the *Crimson*. And so Lewontin wrote, sharing the submitted note with Shockley himself, "Shockley is a racist ideologue who is abysmally ignorant of genetics: Agitators like George Wallace and the head of the American Nazi party do not have academic credentials and attempt to address a mass audience of disaffected people, appealing directly and crudely to their passions and fears. Shockley, trading on his academic standing in electronics, and addressing himself primarily to educated and generally satisfied elites, attempts to propagate his ideology by the use of fallacious pseudo-scientific arguments, while drawing around himself a cloak of false academic objectivity. That is the iota of difference between them. Shockley should be totally ignored." Seizing a final chance to mock and provoke Shockley, Lewontin concludes his letter saying, "I do not know whether this meets your objections to my former remarks. At any rate it has the advantage of being more precise."[78]

Lewontin's involvement in scientific matters related to race was fitting given his role as one of the world's preeminent authorities on human genetic diversity and evolutionary biology, as well as a pioneer in molecular genetics. Lewontin's 1966 articles on genetic variation, written with his colleague Jack Hubby, helped to revolutionize the use of molecular technologies (in this case gel electrophoresis) in the study of evolution.[79] Because of this work, evolutionary biologists were finally able to directly study the amount of genetic diversity in populations by examining proteins.[80] In 1972 Lewontin published a groundbreaking article on this subject called "The Apportionment of Human Diversity," which today remains a classic accounting of human diversity that soundly rejected human racial classification as having "no social value" and "virtually no genetic or taxonomic significance."[81] Unlike his mentor, Dobzhansky, who throughout his career struggled to hold on to the population genetics notion of race that he himself had helped to develop, Lewontin rejected both a social and biological race concept, and he did so based on genetic data revealed with technologies unavailable to scientists like Dobzhansky just a decade earlier. Lewontin challenged a traditional population genetics definition of race by showing that populations were more genetically diverse than once thought; indeed, most genetic variation (85.4 percent) was "contained within" racial groups or "between populations within a race" (8.3 percent), whereas only 6.3 percent of "human variation was accounted for by racial classification."[82] These findings, it should be mentioned, have been confirmed by more recent genetic studies utilizing DNA sequencing data as opposed to Lewontin's immunological data.[83] Aided by a technological breakthrough that he himself had shepherded into evolutionary biology (the use of gel electrophoresis in population genetics) and driven by his mentor's antiracist zeal and his own sense of the relationship between science and politics, Lewontin carried out what was, in a purely scientific context, the most direct proof that, biologically, the category of race in humans was obsolete. In *Sociobiology* Wilson acknowledged as much, quoting "The Apportionment of Human Diversity" to make the point that a similar type of distribution of frequency of behavioral traits could be expected in populations as well.[84]

At the end of the twentieth century, the geneticist L. Luca Cavalli-Sforza confirmed Lewontin's findings using contemporary DNA techniques. His results showed that there was no significant genetic discontinuity between any so-called races in our species that would justify the use of racial classification in humans. Cavalli-Sforza believed that these results and the results of other studies implied that population genetics and evolutionary biology had satisfactorily shown that the "subdivision of the human population into a small number of clearly distinct, racial or continental, groups . . . is not supported by the present analysis of DNA." Given that studies had confirmed Lewontin's results for almost three decades, Cavalli-Sforza believed that "the burden of proof is now on the supporters of a biological basis for human racial classification."[85]

Just a year before the publication of *Sociobiology*, Lewontin published his seminal work *The Genetic Basis of Evolutionary Change*, which was an examination of the nature of genetic diversity among organisms and the significance of that diversity in organismic evolution.[86] As with Dobzhansky's book *Genetics and the Origin of Species*, Lewontin's book also began as a Jessup Lecture at Columbia University. Writing in the

*Quarterly Review of Biology*, Marcus Feldman called *The Genetic Basis of Evolutionary Change* "the most important population genetics book in 10 years."[87] Yale University biologist Jeffrey Powell called Lewontin's book a "classic."[88] Even the philosopher of science Michael Ruse, who would later collaborate with Wilson on books about sociobiology and ethics, wrote that "any philosophical debates about the nature of biology, specifically evolutionary biology, must take account of this work, if not begin with it." In an observation that tied teacher to student, Ruse acknowledged that Lewontin's work was the most significant step in the furtherance of the debates over the nature of evolutionary theory "since Dobzhansky first published his *Genetics and the Origin of Species.*"[89]

That Wilson and Lewontin were heading for an intellectual clash could certainly have been forecast by anyone familiar with their respective works. As the sociologist Ullica Segerstråle points out, readers of *The Genetic Basis of Evolutionary Change* would have recognized that Wilson's *Sociobiology* was a ripe target for Lewontin. In his book, Lewontin was critical of the current state of population genetics and "suggested that fundamental theoretical revisions would have to be made within that field for it to produce valid predictive statements," most notably the erroneous idea that the relationship between genes and traits was "akin to separate beans within a bag."[90] Segerstråle notes that Wilson was aware of this critique, and in *Sociobiology* he "approvingly discussed exactly Lewontin's points," writing, "It remains for sociobiology to exploit both levels as opportunity provides," meaning that he advocated provisionally utilizing existing models and formulas and would wait for new ones to arise.[91]

Looking back on the critiques of *Sociobiology* more than ten years after the intensity of the debate had waned, Wilson still asserted that the nature of those critiques were politically, not scientifically, motivated. "The attacks were often personal in nature," he wrote in a brief autobiographic article published in the 1980s. "The critics implied that I and others working in this area were promoting racism, sexism, and other political evils, either deliberately or else as a side product of being enculturated by a capitalist-imperialist state." And in his view, he had been vindicated: "Fortunately, few people in the academic community believed the charges of Science for the People and INCAR, especially when it became more apparent that my colleagues in the radical Left were promoting a political philosophy and not just defending society from genetic determinism." Yet, curiously, even as he himself criticized the political nature of the sociobiology debates, he acknowledged that "as intellectuals and the American public at large shifted more toward conservatism in the late 1970s, the purely ideological opposition to human sociobiology diminished to near insignificance."[92]

Unlike its reductionist predecessors, sociobiology did not speak in bold tones of genetically driven racial hierarchies. Sociobiology was different; instead of proclaiming racial inferiority for certain groups, it provided a framework to naturalize racism by offering a genetic rationale for it. The tenets of sociobiology—by rooting a genetic basis for complex human social behavior in Darwinian natural selection and Mendelian genetics (the evolutionary or modern synthesis)—helped reintroduce the idea that human difference generated biological impulses embedded in our genomes. Whereas Dobzhansky tried to eliminate typology from the biological race concept, sociobiology reintroduced it by rooting the perception and meaning of difference (e.g., xenophobia and ethnocentrism) in our genes.

# 11

# RACE IN THE GENOMIC AGE

The 1953 discovery of the structure of deoxyribonucleic acid (DNA) by Francis Crick and James Watson ushered in a new era in the biological sciences. Just at the moment that the evolutionary synthesis had become the core theoretical construct in biology, a revolution in molecular biology was quickly emerging. Watson and Crick discovered DNA's double-helical structure by building on discoveries by biochemists including Oswald Avery, Maclyn McCarty, and Colin MacLeod (their work showed that nucleic acids, not proteins, constituted genes),[1] Erwin Chargaff (his greatest find—that in all organisms the ratio between the nucleic acids adenine and thymine to guanine and cytosine is always 1:1),[2] and Rosalind Franklin[3] and Maurice Wilkins (whose work in X-ray crystallography helped conceptualize the double helical structure[4] of DNA).[5]

Until the 1960s, molecular and evolutionary biology developed largely independently of one another. Molecular and evolutionary biologists were trained in different scientific traditions (an emphasis on biochemistry versus organismal biology) with different visions of what biology should be. By the early 1960s these divisions hardened, even between colleagues in the same department. For example, at Harvard in the early 1960s, according to historian Michael Dietrich, James Watson, then chair of Organismic and Evolutionary Biology, "helped polarize the department into what he believed were those working on the 'cutting edge' of biology in molecular biology and those languishing with the concerns of 'classical' biology such as evolution and systematics." The divisions between the two groups of biologists ran deep. The evolutionary biologist Theodosius Dobzhansky once called molecular biology a "glamour field" that was "intellectually shallow."

But Dobzhansky, drawing on his own intellectual legacy as a unifier, sought to reconcile these two distinct, "though overlapping and complementary," approaches to biological research. Dobzhansky proposed a compromise that sought to understand both approaches as "essential for understanding the unity and diversity of life at all levels of interpretation." On the one hand, molecular biology was shaped by a Cartesian or reductionist approach, which understood biological phenomena in the context of chemistry and physics. On the other hand, a Darwinian approach sought to understand biological phenomena in terms of the "adaptive usefulness of structures and processes to the whole organism and to the species of which it is part." Ultimately, though, as Dobzhansky famously noted, "nothing makes sense in biology except in the light of evolution."[6]

However, by the mid-1960s, a new generation of evolutionary biologists—including Richard Lewontin and Jack Hubby at the University of Chicago, working on protein electrophoresis, as well as those experimenting with molecular data to make possible "a quantitative estimate of the 'genetic distance' between different species"—fostered significant collaborations between the fields.[7] But throughout the remaining decades of the twentieth century, despite such bridges, the two approaches remained relatively distinct. The molecular biologist Michel Morange argues that, "although both disciplines use the word 'gene,' they have not sought to bring their two meanings closer, or even to confront them. For the molecular biologist, a gene is a fragment of DNA that codes for a protein. For the population geneticist, it is a factor transmitted from generation to generation, which by its variations can confer a selective advantage (positive or negative) on the individuals carrying it."[8]

With the announcement of the Human Genome Project in 1989, it seemed as if molecular biology had prevailed. Genomics, an approach to biology that utilizes molecular technologies to sequence regions of, or an entire complement of, an organism's DNA (known as a genome), promised to revolutionize science and medicine. According to a 1988 National Research Council report, the primary goals of the Human Genome Project were (1) to construct a map and sequence of the human genome; (2) to develop technologies to "make the complete analysis of the human and other genomes feasible" and to use these technologies and discoveries to "make major contributions to many other areas of basic biology and biotechnology"; and (3) to focus on genetic approaches that compare human and nonhuman genomes, which are "essential for interpreting the information in the human genome."[9]

Many in the academic community, including both natural and social scientists, feared that the reductionism so central to both the epistemological and ontological approaches of the Genome Project would reignite a form of biological determinism not seen since the days of social Darwinism and eugenics.[10] Some also feared that racial science would emerge with renewed vigor in the genomic age.[11] To attempt to address these fears at the outset of the Human Genome Project, James Watson called for at least 3 percent of the Genome Project's annual budget to go toward the study of ethical, legal, and social implications of genome research, as well as the formulation of policies to address such issues. Thus was born the Ethical, Legal, and Social Implications (ELSI) Research Program of the Human Genome Project. The bioethicist Eric Juengst hoped that ELSI would "help optimize the benefits to human welfare and opportunity from the new knowledge, and to guard against its misuses."[12] Since its official formation in 1990, ELSI has funded more than $100 million in research grants.

It remains too early in the genomic revolution to tell what the long-term impact of genomics on the state of the biological race concept will be. The announcement of the completion of the human genome's draft sequence at a June 2000 White House ceremony suggests, however, that the race issue is not going away. At that event, President Clinton, flanked by genome sequencers Francis Collins and Craig Venter, announced the completion of a draft sequence of the human genome. Collins, head of the National Human Genome Research Institute, and Venter, then president of Celera Genomics, had fiercely competed to complete separate draft sequences. But in a spirit of cooperation brokered by the White House, the two scientists together offered their genomic gift to the world—one that is enhancing our understanding of human biology and, in turn, helping public health and medical professionals prevent, treat, and cure disease.[13]

On that day, Venter and Collins also emphasized that their draft genome sequences were confirming what many natural and social scientists had been arguing for decades, that human genetic diversity cannot be captured by the race concept and that all humans have genome sequences that are 99.9 percent identical. At the White House ceremony Venter said, "The concept of race has no genetic or scientific basis."[14] A year later, Collins wrote, "Those who wish to draw precise racial boundaries around certain groups will not be able to use science as a legitimate justification."[15] These conclusions were not novel. Research conducted by population and evolutionary biologists, anthropologists, and historians has shown since the 1930s that racial typologies are not good markers of human genetic diversity,[16] that human populations differ from one another genetically in the relative frequency of alleles,[17] and that the concepts of race and racism have identifiable and changing histories.[18]

Yet since the completion of the draft sequences and the statements on race by Venter and Collins, many still hold fast to the belief that race is, in fact, a biologically meaningful classification.[19] There have been two general approaches to justifying use of the race concept in genomics. One is exemplified by Neil Risch, a University of California, San Francisco, statistical geneticist and genetic epidemiologist. Risch believes that "identifying genetic differences between races and ethnic groups . . . is scientifically appropriate." He argues that race is essential to help determine "differences in treatment response or disease prevalence between racial/ethnic groups" and strongly supports the "search for candidate genes that contribute both to disease susceptibility and treatment response, both within and across racial/ethnic groups."[20] Such an approach exemplifies scientists who utilize the race concept in genomic research and who claim that technological and methodological improvements allow them to examine human diversity with increasing precision that is disconnected from any social prejudices about human difference. Critical of this approach are natural and social scientists who insist that the race concept is a flawed, inaccurate way to measure human genetic diversity that is inseparable from social prejudices about human difference.

A second and more common approach to the race concept among genome scientists is to conflate these two seemingly contradictory viewpoints. And this is the very paradox of the genomic age when it comes to race: scientists say that race does not accurately capture human genetic diversity, yet at the same time, some of those same scientists claim that race is a useful proxy to best capture that genetic diversity—a proxy that is especially useful in clinical settings. By 2005, for example, Francis Collins had shifted from criticisms of the race concept to advocating the need to study how genetic variation and disease risk are correlated with what he called "self-identified race, and how we can use that correlation to reduce the risk of people getting sick."[21] This paradox is embedded in the practice of genomics in the twenty-first century. Self-reported racial identity remains an essential variable used at all stages of genetic research.[22]

The first major controversy related to the race concept and the Human Genome Project was the Human Genome Diversity Project (HGDP), which was proposed in the early 1990s by leading evolutionary biologists and population geneticists as a "resource that is aimed at promoting worldwide research on human genetic diversity, with the ultimate goal of understanding how and when patterns of diversity formed." Project organizers believed that the genetic information garnered from it would likely "prove useful to several areas of biomedical research." Data from the HGDP, it was hoped, could help estimate the incidence of recessive genetic diseases around the world, help identify the genetic variants that contribute to disease, and examine the "contributions of environmental factors to complex human disease."[23]

In order to accomplish its goals, the project collected DNA samples from thousands of populations across the globe, including the world's remaining indigenous populations. And it was these sampling methods and the language used to describe indigenous groups that fueled opposition to the project in some corners of academia and among indigenous rights groups. Even though project organizers were decidedly antiracist—early on, some project scientists sought to sever ties with the historic use of the race concept in scientific studies of human populations by abandoning the category of race in their analysis, proposing to use categories of *group* and *population* instead[24]—critics accused HGDP scientists of racism and colonialism. Non-HGDP scientists were openly critical of the organization of the project, particularly of its framework for studying human populations. Although the HGDP saw as part of its mission to improve understanding of the diversity of non-Europeans (so as not to fall into the trap of seeing Europeans as genetically diverse and non-European groups as genetically homogeneous), the use of terms like "tribe" and "indigenous group" by project scientists left critics, including anthropologist Alan Swedlund, worried that the project might promote racism. Swedlund called the project "21st-century technology applied to nineteenth-century biology."[25] At the same time, indigenous groups worried that they were perceived as human fossils that needed to have their DNA sampled before they disappeared.[26]

The problem, as the sociologist of science Jenny Reardon has argued, is that the population genetics–based race concept can be both a social and a scientific idea, and that therefore not all HGDP scientists "shared the same understanding of racial categories." Some "believed that scientists could continue to use racial categories as long as they properly limited their use" to scientific and medical research. Some project scientists accepted race only as a sociocultural concept. Other scientists advocated using traditional racial categories in conjunction with a more specific understanding of the population under study. And, finally, some participating scientists held contradictory views about the race concept, using it in some contexts and not in others.[27]

More recently, in 2002 the International HapMap Project set out to "determine the common patterns of DNA sequence variation in the human genome" in order to develop "a map of these patterns across the genome." The map, using DNA samples from "populations with ancestry from parts of Africa, Asia and Europe," would be used to determine "the genotypes of one million or more sequence variants, their frequencies and the degree of association between them." The HapMap Project, organizers hope, "will allow the discovery of sequence variants that affect common disease, will facilitate development of diagnostic tools, and will enhance our ability to choose targets for therapeutic intervention."[28]

HapMap organizers have insisted, much like HGDP scientists did, that the project is not about measuring racial differences in the hope of uncovering disease-related information (and then associating certain racial groups with certain diseases) but rather represents the belief that human genetic variation is key to understanding the distribution of disease across human populations.[29] The HapMap's Web site, for example, asserts that "the information emerging from the Project is helping to demonstrate that common ideas about race emerge largely from social and cultural interactions and are only loosely connected to biological ancestry."[30] The website's "Guidelines for Referring to the HapMap Populations in Publications and Presentations" even warns that describing study populations "in terms that are too broad could result in inappropriate over-generalization." This may, in turn, "erroneously lead those who interpret HapMap data to equate geography (the basis on which populations were defined for the HapMap) with race (an imprecise and mostly socially constructed category)."[31]

Still, despite these calculated efforts to separate human genetic diversity research from past racial science, some HapMap critics believe that the project risks recapitulating typological approaches to race by relying on geographic ancestries—the very European, Asian, and African categories that are central to the data-collection methods of the HapMap. As legal scholar Jennifer Hamilton has pointed out, such "taxonomies of geographical ancestry reflect familiar divisions that map rather neatly onto earlier racial taxonomies (e.g., Negro, Caucasian, Mongol; Africanus, Europeanus, Asiaticus)."[32]

The emerging area of personalized medicine provides a third example of the intersection of race and science in the genomic age. This new field claims that the best treatments are individualized ones based on an individual's genome. A goal of personalized medicine is that people will have their own genomes sequenced, and that an analysis of that data will provide doctors with information both about disease risk and about an individual's pharmacogenomic profile (how one's genes influence responses to drugs). We know, for example, that a group of genes known as cytochrome P450 (the CYP family of genes) play an important role in the metabolism of most clinically used drugs.[33] We also know that differences, or polymorphisms, in the sequence of these genes can alter the clinical responses to drugs. Some of these variations can lead to toxic reactions, while others can impact a drug's efficacy. Identifying these differences is thus a clinically useful and sometimes lifesaving tool. The challenge, however, lies in determining individual pharmacogenomic profiles, and this is where the race concept again intersects with the genomic age. Because we do not yet have the technology where this can occur (in most cases) rapidly and economically, researchers have been turning to racial and ethnic profiles as a proxy for estimating individual risks. For example, the CYP2C9 allele (or gene variant), which mediates metabolism of the anticoagulant drug Warfarin, shows up in higher frequency in what one study refers to as Black and Caucasian populations but is extremely rare in

East Asian populations.[34] The CYP2C9 pharmacogenomic literature is rife
with such studies examining alleles in racial, ethnic, and national groups,
including, for example, "Swedes," "African Americans," "Han Chinese,"
and "inner-city Hispanics."[35]

As Dobzhansky has taught us, populations have varying allele frequencies in all their genes. That concept was the basis of the shift from
a typological (or fixed) understanding of racial differences to an evolutionary synthesis understanding of racial difference, which understood
population-level differences in terms of variation of allele frequencies
between racial groups. Remember also that Dobzhansky argued that how
we choose to arrange these groups was a methodological decision rather
than a reflection of the natural order of things. This basic fact of population and evolutionary biology has led some to be highly critical of the
race-based approach to personalized medicine.

To highlight the complexity of gene frequencies being dynamic both
within and between populations, Craig Venter and his colleagues used the
published gene sequences of two self-identified Caucasian men—himself
and James Watson—to show how two white individuals could have very
different pharmacogenomic profiles. Venter, for example, is an "extensive
metabolizer" of CYP2D6, known to be involved in metabolizing codeine,
antipsychotics, and antidepressants. Watson, on the other hand, has a variant of that gene that makes him an "intermediate metabolizer" of this class of
drugs. From a race-based pharmacogenomics perspective, Venter's genetic
profile was predictable. Watson's allele, however, is rare among self-identified
Caucasians but seen in high frequency in East Asian populations. If one
were to have used race alone to assess his ability to metabolize this class of
drugs, Watson would not have received the proper care. Venter and his colleagues argue that this and other examples illustrate why the best choice is
looking directly at an individual's "genomic sequence instead of relying on
a patient's appearance or self-identified ethnicity." Venter and his team also
point out that gene variants do not always behave as expected: "Although
these complications may be due to other genetic differences, cultural factors
such as diet and environment can also influence drug response."[36]

What is interesting in the context of the history of the race concept,
however, is how these examples—the HGDP, the HapMap Project, and
pharmacogenomics—illustrate similar patterns of racial research in the

genomic age. That is, that scientists readily claim the race concept is both a reasonable proxy for genetic diversity even as they recognize its limited utility as a classificatory tool. More recently, two meetings sponsored by the National Human Genome Research Institute in 2007 and 2008 sought to address the current challenges of studying human genetic diversity, particularly given the burgeoning interest in the relationship between self-identified race and health disparities. At the 2007 conference, "Frontiers in Population Genomics Research Meeting," one speaker called for scientists to "eliminate reliance on the construct of race" and for population geneticists to "engage populations that adequately reflect human diversity."[37] The 2008 meeting, "Understanding the Role of Genomics in Health Disparities: Toward a New Research Agenda," was jointly sponsored by the National Human Genome Research Institute, the National Cancer Institute, and the National Center on Minority Health and Health Disparities of the National Institutes of Health (NIH). Attendees, representing many NIH institutes as well as natural and social scientists with a research interest in disparities, made several recommendations based on two days of lectures and discussions.

In an apparent attempt to, at the very least, address how the race concept is used in scientific studies, the group recommended that NIH Funding Opportunity Announcements "should require that investigators justify their use of racial and ethnic categories relative to the questions that they are asking and the methods they are using in particular research." The group also recommended the development of "statistical tools to analyze complex populations as a single unit based on genomic/molecular profile (or characteristics) rather than stratify by race and ethnicity." The report remains unpublished and it is unclear what, if any, impact its recommendations will have.[38] Despite the best intentions of such reports, it has been over fifty years since Dobzhansky issued his challenge to the field in 1962—he said then that "the problem that now faces the science of man is how to devise better methods for further observations that will give more meaningful results"— and we are still struggling with the meaning of race in biology.[39]

A more immediate problem is that for now, despite some trying to find useful solutions to these challenges, the NIH still reifies racial categories in its grant applications; scientists are required to describe recruitment strategies for human subjects that emphasize traditional racial classifications.

Scientists working at or funded by the NIH are mandated to report race based on the U.S. Census categories and following the standards set forth by the White House Office of Management and Budget Directive No. 15, which "defines minimum standards for maintaining, collecting and presenting data on race and ethnicity for all Federal reporting."[40] The "Targeted/Planned Enrollment Table," a fixture in all NIH grant applications that include human subjects, divides humanity into five racial groups: "American Indian/Alaska Native," "Asian," "Native Hawaiian or Other Pacific Island," "Black or African American," and "White." Although the form was recently updated to include categories for "more than one race" and "unknown or not reported," it still reinforces the most antiquated notions of this fundamentally flawed concept, making it more difficult to explore the subtlety and complications inherent to human genetic diversity.[41] So long as the NIH is obligated to follow Office of Management and Budget rules, the system will perpetuate this racial paradigm.

While the NIH meetings were a call to action for the field, some have more directly attempted to address Dobzhansky's challenge to devise better methods to understand human diversity. The molecular epidemiologist Timothy Rebbeck and bioethicist Pamela Sankar at the University of Pennsylvania, for example, acknowledge that "much of the persistent controversy over the use of the terms 'ethnicity,' 'ancestry,' or 'race' may be attributable to the imprecision of their use." They believe that defining the terminology of race on a study-by-study basis may be a way to simultaneously embrace the inconsistencies of the race concept and find meaning in its specific context. Yet ultimately, they recognize that as a measure, race "may have utility in increasing study efficiency or reducing confounding," but "an important future direction for research will be to develop new measures that correlate with SIRE [self-identified race or ethnicity] that may better reflect the complex nature of this variable."[42]

If these struggles over race at the outset of the genomic age can teach us anything about the history of the race concept, it is that even a generation of scientists reared in the wake of the evolutionary synthesis and population genetics (who had been trained to reject typology as a component of their analysis of populations) still struggle with utilizing a concept that has such a contradictory and sometimes awful history. These scientists—contradictions and all—indeed reflect and are working in the traditions of Dobzhansky and the evolutionary synthesis.

# EPILOGUE

———

## Dobzhansky's Paradox and the Future of Racial Research

I n 1998 the public health psychiatrist Mindy Fullilove published an article in the *American Journal of Public Health* challenging the utility of race as a variable in health research, pointing to the risks of relying on an archaic and imprecise way to organize human diversity and calling upon the field "to abandon race as a variable in public health research." The race concept "is an arbitrary system of visual classification that does not demarcate distinct subspecies of the human population," Fullilove asserted, arguing that the concept could not "provide the information we need to resolve the health problems of populations." In its place, she proposed that public health professionals "invent a new science that embodies the human rights and civil rights essential to the health of human populations." Although Fullilove acknowledged that " 'race' is an ingrained part of personal identity," she believed that abandoning "racial classification schemes is to challenge deeply held and socially endorsed ways of seeing one's place in the world." Fullilove's proposition was as much a call to science to find new methods to study the relationship between health outcomes and human diversity as it was to remind her fellow scientists that "following the illusion of race cannot provide the information we need to resolve the health problems of populations."[1]

Fullilove was not, of course, the first scientist or scholar to call for rethinking the scientific methodologies that measure human diversity. At the turn of the twentieth century, W. E. B. Du Bois called into question the scientific legitimacy of the biological race concept in *The Health and Physique of the Negro American.*[2] In the early 1940s the anthropologist Ashley Montagu called for an abandonment of the race concept in his classic work *Man's Most Dangerous Myth*.[3] The evolutionary biologist Theodosius Dobzhansky, who had in the 1930s and 1940s been largely responsible for reimagining and preserving the race concept in biology in his work on the evolutionary synthesis, came to believe by the 1960s that "the problem that now faces the science of man is how to devise better methods for further observations that will give more meaningful results."[4] And in 1972, the evolutionary biologist Richard Lewontin wrote that human racial classifications have "virtually no genetic or taxonomic significance."[5] Despite these calls, the population genetics–based race concept as first articulated by Dobzhansky is alive and well in the biological sciences.

Part of the challenge of situating this history in present-day debates about race is that, in general, unlike pre-twenty-first-century arguments about the race concept in biological thought, today's debates generally lack the clear ideological and political antipodes of racist and nonracist, reactionary and liberal. Instead, today's scientific debates take place (with some exceptions, of course) in a way that seeks to avoid the ideological and racist baggage of the past. Claims for race-based medicine in the genomic era are allegedly about drug safety and efficacy, reducing health disparities, and ultimately bettering human health and saving lives.[6] Most of the discussion today about race in biology concerns the noble goal of utilizing an individual's self-identified ancestry to help determine the best course of medical treatments as we move closer to personalized medicine. Although there is a legitimate fear that this type of research will serve only to reify racial categories and harden racial prejudices, that is not, in most cases, its purpose. The question as it is often framed in today's debates about race is not whether racial classifications are harmful—and this is with an acknowledgment that race is a social construction—but whether race is a valuable biological variable and whether we can best improve the health of individuals and populations by continuing to use this variable.

In public health, the attention to health disparities has further deepened the interest of public health professionals in the race concept. Some in public health have chosen to integrate a socially constructed race concept into novel methodologies designed to study health disparities, while others have drawn on quasi-typological ideas of race to understand the relationship between biology and health. The work of Nancy Krieger at Harvard, for example, rejects the biological race concept but utilizes race in a novel methodology that she calls the "Public Health Disparities Geocoding Project." Her method seeks to understand the underpinning of racial disparities in health by examining the underlying socioeconomic causes of disparities. In order to overcome a virtual "absence of socioeconomic data in most U.S. public health surveillance systems," she has developed a method that utilizes geocoded residential addresses and area-based socioeconomic measures.[7] Others have made alternative proposals. Ichiro Kawachi, Norman Daniels, and Dean Robinson argue that "historical, political, and ideological obstacles" have "hindered the analysis of race and class as codeterminants of disparities in health." They propose that "racial and class identities are mutually constitutive," as race is "neither a biologically meaningful category nor a proxy for class, but is a separate construct from class, more akin to caste." They call for a strengthening of the U.S. data infrastructure "to improve the measurement of race/ethnicity as well as class/socioeconomic status" so that policy makers can better "monitor the effects of their policies on health disparities."[8] Still others have called for more clarity in the use of race in health research. "At a minimum," R. Dawn Comstock, Edward Castillo, and Suzanne Lindsay argue, "researchers should clearly state the context in which these valuable epidemiologic and public health research variables are being used, describe the method used to assess these variables, and discuss all scientific findings." Doing so, they believe, "will ensure continued constructive dialog about the interpretation of findings regarding race or ethnicity."[9]

Finally, epidemiologist Camara Jones argues that racism, not race, may be a "root cause of observed race-associated differences in health outcomes." Race, according to Jones, "is a social classification in our race-conscious society that conditions most aspects of our daily life experiences and results in profound differences in life chances." An understanding of what she calls the three levels of racism—institutionalized racism ("differential access to the goods, services, and opportunities of society by race"), personally mediated racism ("prejudice and discrimination, where prejudice means differential assumptions about the abilities, motives, and intentions of others according to their race, and discrimination means differential actions toward others according to their race"), and internalized racism ("acceptance by members of the stigmatized races of negative messages about their own abilities and intrinsic worth")—may offer a better understanding of the etiology of health disparities as well as concrete strategies to redress them.[10]

The latest ways to utilize race and begin to rethink how to create a measurement variable from human genetic diversity are, despite their best intentions, overwhelmed in the larger research enterprise by studies that claim that race is, in fact, a methodologically sound classificatory tool in the biological sciences.[11] Continued claims that race is, in fact, a useful tool for classification is driven by four factors. First, the reductionist ontology that underlies genomic science has given rise to the belief that biology is destiny, and that both at the individual and population levels, genes hold the ultimate information for understanding our health and understanding who we are. The philosopher of science Sahotra Sarkar worries that such reductionism in genetics "attempts to explain phenotypic properties from a genotypic basis without attributing any particular structure to the genotype."[12] Claims that race is a meaningful biological classificatory tool do make sense as an extension of reductionism as the line between phenotype and genotype is blurred.

Second, genomic technology has enhanced our ability to examine the 0.1 percent of nucleic acids in the human genome that, on average, vary between individuals. Over the past decade, the Human Genome Project and the technological revolution that spawned it are driving novel research to examine these differences. Many of these 0.1 percent differences will involve phenotypic variations including hair color, body shape and size, skin tone, disease susceptibility, and blood type. The 0.1 percent of variations between people can also be used to study both individual and population-level traits, including disease risk. However, this is still prohibitively expensive to do for everyone, so scientists are instead using population and ethnic groups as well as races as proxies to measure susceptibility and risk. One area where we are seeing some of the earliest manifestations of such an approach is in pharmacogenomic testing—a way to predict, based on an individual's genotype, how one will react (either negatively or positively) to a particular drug. Will drug A, for example, be an effective and safe treatment for patient Y, or will it be toxic and nonefficacious? But, for the moment, because of the expense of sequencing individual genomes, scientific studies are often relying on an individual's race to make pharmacogenomic predictions.

The problem with such an approach, as the bioethicist Sandra Soo-Jin Lee points out, is that it inevitably "suggests that the population under study is assumed to be homogeneous with respect to allele frequencies but in fact comprises subpopulations that have different allele frequencies for the candidate gene."[13] In other words, as Dobzhansky tried to teach us many years ago, "racial differences are more commonly due to variations in the relative frequencies of genes in different parts of the species population than to an absolute lack of certain genes in some groups and their complete homozygosis in others. . . . *Individuals carrying or not carrying a certain gene may sometimes be found in many distinct races of a species.*"[14] It is therefore not accurate to make assumptions about an individual's genes based on their race. Race as a proxy predictions will sometimes be accurate if the gene frequencies for, let's say, a drug's metabolism are high enough in a particular racial group. But inevitably, many will receive no benefit or may even be harmed based on these assumptions.

The 0.1 percent difference between people is also being used to promote a closer examination of genetic ancestry. A simple Google search reveals companies that claim a pretty simple correlation between your genes and your race. Although it doesn't use the term "race," one company allows you to "discover your estimated percentage of ancestry from four different population groups: European, Indigenous American, Sub-Saharan African, and East Asian."[15] Another company's promotional video claims that DNA "is the key to discovering where your family is from, and learning about the places and culture that make you who you are."[16] Critics, however, have pointed out that these tests "cannot pinpoint the place of origin or social affiliation of even one ancestor with exact certainty."[17] Keith Wailoo, Alondra Nelson, and Catherine Lee also worry that "the mere hint that genetic markers are distributed in different frequencies across populations has led some people to quickly treat such variation as a proxy for racial and ethnic differences, lending renewed authority to biological conceptions of human difference."[18]

Third, the critical task of understanding and reducing health disparities has researchers looking at all possible explanations, including genetic ones, for disparities in health outcomes. The renewed focus on race and genetics suggests, however, that an analysis of the complex relationship between individuals, populations, and health has been surrendered to a simplistic, racialized worldview. An inability to digest (and frequently even to acknowledge) these complexities restricts scientific theory and practice to simplicity when complexity is needed. This underlies the drive to correlate race, genetics, and health disparities.

Fourth and finally, the history of the biological race concept demonstrates that race is deeply embedded in scientific and social thought, and that racialized thinking was an integral part of genetics in the twentieth century.[19] This history has shaped scientific thinking about human difference as well as popular thinking about that difference. That Craig Venter's criticisms of the race concept at the outset of the twenty-first century are similar to those made by W. E. B. Du Bois at the beginning of the twentieth—calling into question both the utility of racial classifications and the claim that health disparities can be explained by elucidating the relationship between genetics and race—suggests that we are having frustratingly similar arguments about race and human difference despite the benefit of one hundred years of knowing better.[20]

In light of this history, we should not underestimate the tenacity of the race concept. As scientists continue to struggle with the meanings of race, they should remember Dobzhansky's paradoxical thinking about the race concept: on the one hand, Dobzhansky believed the race concept to be a critical methodological tool for biologists to make sense of genetic diversity within species.[21] On the other hand, Dobzhansky understood the imprecise nature of the term, its limited utility, and its potential for abuse. But despite these acknowledged contradictions, Dobzhansky held fast to and publically defended the concept for much of his career. Toward the end of his career he struggled to disown this paradox. He did this most probably because of the ways in which he saw the biological race concept being appropriated in the early years of the civil rights movement by those fanatically opposed to African American civil rights, because of disputes with other scientists who he believed were misusing the race concept in their work, *and ultimately* because he seemed to realize that he was fighting a Sisyphean battle in trying to separate what he had formerly considered the separate domains of the scientific and social meanings of the biological race concept. Nevertheless, genetics continues to function within Dobzhansky's paradox. Despite his brilliance as an evolutionary theorist, Dobzhansky left us with only a warning and not a new method to study human diversity. History has shown that even acknowledging that race has a social meaning cannot disconnect the concept from its typological and racist past (or present). Despite the best intentions of scientists and scholars, race will always remain what Ashley Montagu once called a "trigger word. . . . Utter it and a whole series of emotionally conditioned responses follow."[22]

And so while the race concept can provide us with some information about health and human evolutionary history, ultimately that information is incomplete. What we know from Dobzhansky, his sometimes collaborator L. C. Dunn, the evolutionary geneticist Richard Lewontin, and, most recently, Craig Venter, is that race is not a particularly useful measure of human genetic diversity. We also know from Du Bois and other historians and social scientists that the biological race concept is inseparable from its social history, and that race can be harmful to an understanding of both human health and evolutionary history.

We are a genetically diverse species, and there is meaning in that diversity. But as a species we seem thus far unable to reliably distinguish between the scientific ramifications and the social meanings of human difference. "Race" is a historical, not a scientific, term. Yet until the scourge of racism is eliminated from our lives and institutions, developing scientific methods unburdened by racial ideology to study human difference will be limited by the historical nature of the race concept itself.

In 1940, toward the end of his career, W. E. B. Du Bois penned his autobiography, *Dusk of Dawn: An Essay Toward an Autobiography of a Race Concept*. David Levering Lewis, Du Bois's biographer, believes that the book was intended to be "not so much the story of his own life, then, but the autobiography of the twentieth century as it had lived through him."[23] The book is thus, in large part, both a personal account of the meaning of race in Du Bois's own life and an evaluation of its political, sociological, and scientific meanings over the course of his lifetime. Du Bois understood that scientific concepts of race had played a critical role in both providing support for racist claims and in buttressing the racial order, writing in *Dusk of Dawn* that he "had too often seen science made the slave of caste and race hate." Yet Du Bois also understood that the world and the race concept had changed, in some ways for the better and others for the worse, over his lifetime. "Life has its pain and evil—its bitter disappointments," Du Bois wrote at the book's close, "but I like a good novel, and in healthful length of days, there is infinite joy in seeing the World, the most interesting of continued stories, unfold, even though one misses THE END."[24]

In our own time, we continue to struggle with the power of the race concept. Biologists may today develop new ways of studying human populations that, to whatever degree, distance themselves from the race concept and its historical baggage, and those new methods may, in fact, be an improvement over where we stood before. Yet we should not be waiting for or expecting science and scientists to change our thinking about race. Science may have helped bring us to this point, but it is unlikely to extract us from it.

That social and natural scientists have been rejecting, abandoning, and discrediting the race concept for over a century suggests that, for now, the race concept in biology is here to stay.

# NOTES

## INTRODUCTION

1. Ashley Montagu, *Man's Most Dangerous Myth: The Fallacy of Race* (Walnut Creek, Calif.: AltaMira Press, 1997); Jacques Barzun, *Race: A Study in Superstition* (New York: Harper and Row, 1965); Barbara J. Fields, "Ideology and Race in American History," in *Region, Race, and Reconstruction: Essays in Honor of C. Vann Woodward*, ed. J. Morgan Kousser and James M. McPherson, 143–77 (New York: Oxford University Press, 1982).
2. Bruce Dain, *A Hideous Monster of the Mind: American Race Theory in the Early Republic* (Cambridge, Mass.: Harvard University Press, 2002), vii.
3. Daniel J. Kevles, *In the Name of Eugenics: Genetics and the Uses of Human Heredity* (Cambridge, Mass.: Harvard University Press, 1995).
4. Karen E. Fields and Barbara J. Fields, *Racecraft: The Soul of Inequality in American Life* (New York: Verso, 2012), 18–19.
5. Ibid., 5–6.
6. Thomas Jefferson, *Notes on the State of Virginia*, ed. William Peden (Chapel Hill: University of North Carolina Press, 1955), 138–140, 143.
7. Ashley Montagu, *Statement on Race* (New York: Oxford University Press, 1972), 9–13.
8. Jenny Reardon, *Race to the Finish: Identity and Governance in an Age of Genomics* (Princeton: Princeton University Press, 2005), 22–23.
9. Ibid.; Lee D. Baker, *From Savage to Negro: Anthropology and the Construction of Race, 1896–1954* (Berkeley: University of California Press, 1998); Gregory Dorr, *Segregation's Science: Eugenics and Society in Virginia* (Charlottesville: University of Virginia Press, 2008).
10. Theodosius Dobzhansky, *Mankind Evolving: The Evolution of the Human Species* (New Haven: Yale University Press, 1962), 266.
11. Ibid., 252–53.
12. Gerald Markowitz and David Rosner, *Children, Race, and Power: Kenneth and Mamie Clark's Northside Center* (Charlottesville: University Press of Virginia, 1996), 90–94.

13. William Stanton, *The Leopard's Spots: Scientific Attitudes Toward Race in America, 1815–59* (Chicago: University Of Chicago Press, 1960), 25.

14. Mae M. Ngai, *Impossible Subjects: Illegal Aliens and the Making of Modern America* (Princeton: Princeton University Press, 2004), 3–5.

15. Baker, *From Savage to Negro*, 3.

16. Charles E. Rosenberg, *No Other Gods: On Science and Society and American Social Thought* (Baltimore: Johns Hopkins University Press, 1961), 1.

17. K. Anthony Appiah and Amy Gutmann, *Color Consciousness: The Political Morality of Race* (Princeton: Princeton University Press, 1996), 41–42.

18. Brian D. Smedley et al., eds., *Unequal Treatment: Confronting Racial and Ethnic Disparities in Health Care* (Washington, D.C.: National Academies Press, 2003); John Yinger, *Closed Doors, Opportunities Lost: The Continuing Costs of Housing Discrimination* (New York: Sage, 1995).

## 1. A EUGENIC FOUNDATION

1. Charles Davenport Papers, American Philosophical Society, Folder, Davenport, Charles B. Lecture: "Racial Traits," February 21, 1921.

2. Daniel J. Kevles, *In the Name of Eugenics: Genetics and the Uses of Human Heredity* (Cambridge, Mass.: Harvard University Press, 1995), 46.

3. Ibid., 85.

4. Mark Pittenger, *American Socialists and Evolutionary Thought, 1870–1920*, History of American Thought and Culture (Madison: University of Wisconsin Press, 1993), 248.

5. Celeste Condit, *The Meanings of the Gene* (Madison: University of Wisconsin Press, 1999), 27; Mark A. Largent, *Breeding Contempt: The History of Coerced Sterilization in the United States* (New Brunswick, N.J.: Rutgers University Press, 2011). 1.

6. Paul Popenoe and Roswell Hill Johnson, *Applied Eugenics* (New York: Macmillan, 1933), 141.

7. William Provine, "Genetics and Race," *American Zoologist* 26 (1986): 857.

8. Ibid., 868.

9. Charles Davenport Papers, American Philosophical Society, Folder, Davenport, Charles B. Lecture: "Do Races Differ in Mental Capacity?" March 1928.

10. Popenoe and Johnson, *Applied Eugenics*, 283–84.

11. Eugenics Record Office Papers, American Philosophical Society, Box 62, Folder A: 974x6#10, "Pedigree of W. E. B. Du Bois"; Mark Aldrich, "Progressive Economists and Scientific Racism: Walter Willcox and Black Americans, 1895–1910," *Phylon* 40 (1979): 1–14.

12. See, for example, Kevles, *In the Name of Eugenics*; Peter Schrag, *Not Fit for Our Society: Nativism and Immigration* (Berkeley: University of California Press, 2010); Stefan Kühl, *The Nazi Connection: Eugenics, American Racism, and German National Socialism* (New York: Oxford University Press, 2002); Laura Briggs, *Reproducing Empire: Race, Sex, Science, and*

*U.S. Imperialism in Puerto Rico* (Berkeley: University of California Press, 2002); Largent, *Breeding Contempt*; Alexandra Minna Stern, *Eugenic Nation: Faults and Frontiers of Better Breeding in Modern America* (Berkeley: University of California Press, 2005).

13. Michael G. Kenny, "Toward a Racial Abyss: Eugenics, Wickliffe Draper, and the Origins of the Pioneer Fund," *Journal of History of the Behavioral Sciences* 38 (summer 2002): 259–83.

14. Kevles, *In the Name of Eugenics*, 3.

15. Francis Galton, *Memories of My Life* (London: Methuen, 1908), 1:141.

16. Karl Pearson, *The Life, Letters, and Labours of Francis Galton, Volume 2: Researches of Middle Life* (Cambridge: Cambridge University Press, 1924), 74.

17. Raymond E. Fancher, "Biographical Origins of Francis Galton's Psychology," *Isis* 74 (June 1983): 228–29.

18. Ibid., 228.

19. Nicholas Wright Gillham, *A Life of Sir Francis Galton: From African Exploration to the Birth of Eugenics* (Oxford: Oxford University Press, 2001), 26–27.

20. Ibid., 30–31.

21. Fancher, "Biographical Origins," 231.

22. Ibid., 232.

23. Biographies and biographical articles on Galton include Gillham, *Sir Francis Galton*; Michael Bulmer, *Francis Galton: Pioneer of Heredity and Biometry* (Baltimore: Johns Hopkins University Press, 2003); Ruth Schwartz Cowan, "Sir Francis Galton and the Study of Heredity in the Nineteenth Century" (Ph.D. diss., Johns Hopkins University, 1969); and Fancher, "Biographical Origins."

24. Kevles, *In the Name of Eugenics*, 6.

25. Galton, *Memories of My Life*, 141.

26. Raymond Fancher, "Francis Galton's African Ethnography and Its Role in the Development of His Psychology," *British Journal for the History of Science* 16 (1983): 67–79.

27. Cowan, "Sir Francis Galton," 25.

28. *Macmillan's Magazine*, 1st Paper, June 1865, 2nd Paper, August 1865, vol. 12:157–66, 318–27.

29. Ibid.

30. Fancher, "Galton's African Ethnography," 67–79; Kevles, *In the Name of Eugenics*, 8.

31. Norma Myers, *Reconstructing the Black Past: Blacks in Britain, 1780–1830* (London: Cass, 1996).

32. See Gretchen Gerzina, *Black Victorians/Black Victoriana* (New Brunswick, N.J.: Rutgers University Press, 2003), and James Walvin, *Black and White: The Negro and English Society, 1555–1945* (London: Lane, 1973).

33. Michael Banton, *White and Coloured: The Behavior of British People Towards Coloured Immigrants* (New Brunswick, N.J.: Rutgers University Press, 1960), 57.

34. Ibid., 58.

35. John Stuart Mill, *Principles of Political Economy* (Toronto: University of Toronto Press, 1965), 319, quoted in Fancher, "Galton's African Ethnography," 68.

36. Cowan, "Sir Francis Galton," 26.

37. Banton, *White and Coloured*, 60.
38. Francis Galton, "Letters of Henry Stanley from Equatorial Africa to the 'Daily Telegraph.' London: 1877," *Edinburgh Review* 147 (January 1878): 177.
39. John C. Kenna, "Sir Francis Galton's Contribution to Anthropology," *Journal of the Royal Anthropological Institute of Great Britain and Ireland* 94 (July–December 1964): 85.
40. Francis Galton, *Hereditary Genius: An Inquiry into Its Laws and Consequences* (Gloucester, Mass.: Smith, 1972), 40.
41. Banton, *White and Coloured*, 59; Warwick Anderson, *The Cultivation of Whiteness: Science, Health, and Racial Destiny in Australia* (New York: Basic Books, 2003).
42. Francis Galton, *Hereditary Genius: An Inquiry into Its Laws and Consequences* (London: Macmillan, 1869), vi.
43. All the review quotes are in Gillham, *Sir Francis Galton*, 170.
44. William B. Provine, *The Origins of Theoretical Population Genetics* (Chicago: University of Chicago Press, 2001), 24.
45. Garland Allen, "Genetics, Eugenics, and Society: Internalists and Externalists in Contemporary History of Science," *Social Studies of Science* 6 (February 1976): 106.
46. Galton, *Hereditary Genius* (1869), 336.
47. Ibid., 336–37.
48. Ibid., 337, 339.
49. Ibid., 338.
50. Ibid., 338–39.
51. Francis Galton, *Inquiries into Human Faculty and Its Development* (New York: Macmillan, 1883), 24.
52. Ibid., 24–25.
53. Ibid., 332.
54. Ibid., 1–2.
55. "Galton's Human Faculty," unsigned review of *Inquiries into Human Faculty and Its Development*, by Francis Galton, *Science* 2 (July 20, 1883): 80.
56. Galton, *Inquiries into Human Faculty*, 305–7.
57. Cowan, "Sir Francis Galton," 144, 200.
58. Ibid., 200.
59. Ibid., 203.
60. Kevles, *In the Name of Eugenics*, xiii.
61. Francis Galton, *Essays in Eugenics* (London: Eugenics Education Society, 1909), 35.
62. Ivan Hannaford, *Race: The History of an Idea in the West* (Washington, D.C.: Woodrow Wilson Center Press, 1996), 123.
63. Audrey Smedley, *Race in North America: Origin and Evolution of a Worldview* (Boulder: Westview Press, 1993), 164.
64. Arthur O. Lovejoy, *The Great Chain of Being: A Study of the History of an Idea* (New York: Harper and Row, 1960), 59.
65. Smedley, *Race in North America*, 164.
66. Ibid., 36–40; Hannaford, *Race*.

67. Smedley, *Race in North America* 165.

68. Georges Louis Leclerc, *Natural History: Containing a Theory of the Earth, a General History of Man, of the Brute Creation, and of Vegetables, Minerals, &c.* (London: Barr, 1792), 317–18.

69. Stephen Jay Gould, *The Mismeasure of Man* (New York: Norton, 1996), 403, 408.

70. Galton, *Inquiries into Human Faculty*, 331.

71. Galton, "Letters," 189.

72. Galton, *Inquiries into Human Faculty*, 310–311.

73. Ibid., 314.

74. Ibid., 316–17.

75. Michael G. Kenny, "Toward a Racial Abyss: Eugenics, Wickliffe Draper, and the Origins of the Pioneer Fund," *Journal of the History of the Behavioral Sciences* 38 (2002): 261.

76. Galton, *Essays in Eugenics*, 25.

77. Susan Lindee, *Moments of Truth in Genetic Medicine* (Baltimore: Johns Hopkins University Press, 2005), 122.

78. Galton, *Inquiries into Human Faculty*, 217.

79. Ibid., 217–18.

## 2. CHARLES DAVENPORT AND THE BIOLOGY OF BLACKNESS

1. Oscar Riddle, "Charles Benedict Davenport," *Science* 99 (June 2, 1944): 441–42.

2. Edwin Black, *War Against the Weak: Eugenics and America's Campaign to Create a Master Race* (New York: Four Walls Eight Windows, 2003), 32–33.

3. Ibid., 33.

4. Daniel J. Kevles, *In the Name of Eugenics: Genetics and the Uses of Human Heredity* (Cambridge, Mass.: Harvard University Press, 1995), 45.

5. Charles Davenport Papers, American Philosophical Society, Lectures H–M, File, Davenport, Charles B. Lecture: "Heredity and Eugenics," 1920.

6. Thomas Kessner, *The Golden Door: Italian and Jewish Immigrant Mobility in New York City, 1880–1915* (New York: Oxford University Press, 1977), 7–8.

7. Osborn to Johnson, May 2, 1924. Henry Fairfield Osborn Collection, File: Johnson; American Museum of Natural History Archives.

8. Osborn to Grant, December 23, 1919. Henry Fairfield Osborn Collection, File: Grant, Madison, Folder 39; American Museum of Natural History Archives.

9. Osborn to Johnson, December 19, 1922. Henry Fairfield Osborn Collection, File: Johnson; American Museum of Natural History Archives.

10. Kevles, *In the Name of Eugenics*, 102–3.

11. William H. Tucker, *The Science and Politics of Racial Research* (Urbana: University of Illinois Press, 1994), 95.

12. Ibid., 94–95; David A. Hollinger, "How Wide the Circle of the 'We'? American Intellectuals and the Problems of the Ethnos since World War II," in *Scientific Authority*

*and Twentieth-Century America*, ed. Ronald G. Walters, 13–31 (Baltimore: Johns Hopkins University Press, 1977).

13. Tucker, *Science And Politics*, 96.

14. Charles Davenport Papers, American Philosophical Society, Davenport to Johnson, December 24, 1923.

15. Charles Davenport Papers, American Philosophical Society, Johnson to Davenport, December 27, 1923.

16. Matthew Jacobson, *Whiteness of a Different Color: European Immigrants and the Alchemy of Race* (Cambridge, Mass.: Harvard University Press, 1998), 93.

17. Mae Ngai, "The Architecture of Race in American Immigration Law: A Reexamination of the Immigration Act of 1924," *Journal of American History* 86 (June 1999): 69–71.

18. Jacobson, *Whiteness*, 95.

19. Ibid., 98.

20. Julie Novkov, "Racial Constructions: The Legal Regulation of Miscegenation in Alabama, 1890–1934," *Law and History Review* 20 (summer 2002): 252.

21. C. Vann Woodward, *The Strange Career of Jim Crow* (New York: Oxford University Press, 1955), 6.

22. Gregory Dorr, *Segregation's Science: Eugenics and Society in Virginia* (Charlottesville: University of Virginia Press, 2008).

23. Charles Davenport Papers, American Philosophical Society, Lectures A–H, File, Davenport, Charles B. Lecture: "A Biologist's View of the Negro Problem," n.d.

24. Kevles, *In the Name of Eugenics*, 46.

25. Charles Davenport Papers, American Philosophical Society, File, Hrdlička, Aleš: Davenport to Hrdlička, November 27, 1906.

26. Charles Davenport Papers, American Philosophical Society, File, Hrdlička, Aleš: Hrdlička to Davenport, November 28, 1906.

27. Gertrude C. Davenport and Charles B. Davenport, "Heredity of Skin Pigmentation in Man," *American Naturalist* 44 (November 1910): 641–72. The second half of the article was published in vol. 44 (December 1910): 705–31.

28. Ibid., (November 1910): 672.

29. William B. Provine, "Geneticists and the Biology of Race Crossing," *Science* 182 (November 23, 1973): 790–96.

30. Davenport and Davenport, "Heredity of Skin Pigmentation," 668.

31. Paul Popenoe and Roswell Hill Johnson, *Applied Eugenics* (New York: Macmillan, 1933), 284, 294.

32. Charles Davenport Papers, American Philosophical Society, File, Popenoe, Paul, Folder 2: Davenport to Popenoe, July 13, 1914.

33. Dorr, *Segregation's Science*, 2–3. See also Harvey Jordan, "The Biological Worth and Social Status of the Mulatto," *Popular Science Monthly* 82 (June 1913): 573–82.

34. Charles Davenport Papers, American Philosophical Society, File, Jordan, H. E.: Jordan to Davenport, July 16, 1913.

35. Charles Davenport Papers, American Philosophical Society, File, Jordan, H. E.: Davenport to Jordan, August 7, 1913.

36. Charles Davenport Papers, American Philosophical Society, File, Jordan, H. E.: Jordan to Davenport, November 27, 1912, and Davenport to Jordan, November 30, 1912.

37. Davenport and Davenport, "Heredity of Skin Pigmentation," 666.

38. Charles B. Davenport, *Heredity of Skin Color in Negro-White Crosses* (Washington, D.C.: Carnegie Institution of Washington, 1913), 28–30.

39. Charles B. Davenport, "The Effects of Race Intermingling," *Proceedings of the American Philosophical Society* 56 (1917): 364.

40. Ibid., 364–68.

41. Peggy Pascoe, "Miscegenation Law, Court Cases, and Ideologies of 'Race' in Twentieth-Century America," *Journal of American History* 83 (June 1996): 44–69; Dorr, *Segregation's Science*.

42. Charles Davenport and Morris Steggerda, *Race Crossing in Jamaica* (Washington, D.C.: Carnegie Institution of Washington, 1929), 477.

43. Lothrop Stoddard, *The Rising Tide of Color Against White World-Supremacy* (New York: Scribner, 1920), 301.

44. Stoddard to Osborn, January 23, 1923. Henry Fairfield Osborn Collection, American Museum of Natural History; Norman Hapgood, "The *New* Threat of the Ku Klux Klan," *Hearst's International* (January 1923): 8–12.

45. Charles Davenport Papers, American Philosophical Society, Folder, Davenport, Charles B. Lecture: "Racial Traits," February 21, 1921.

## 3. EUGENICS IN THE PUBLIC'S EYE

1. Henry Fairfield Osborn, "Address of Welcome," in *Scientific Papers of the Second International Conference of Eugenics*, vol. 1, *Eugenics, Genetics, and the Family* (Baltimore: Williams and Wilkins, 1923), 1.

2. Ibid., 4.

3. Charles Davenport, "Research in Eugenics," in *Eugenics, Genetics, and the Family*, 24.

4. Ibid., 28.

5. William B. Provine, *Sewall Wright and Evolutionary Biology* (Chicago: University of Chicago Press, 1986), 180.

6. Charles Davenport Papers, American Philosophical Society, File, Morgan, Thomas Hunt: Morgan to Davenport, January 18, 1915.

7. Garland Allen, *Thomas Hunt Morgan: The Man and His Science* (Princeton: Princeton University Press, 1978), 228.

8. American Museum of Natural History, *Annual Report of the Trustees of the American Museum of Natural History for the Year (1921)* (New York: American Museum of Natural History, 1921), 31.

9. Ibid., 33.

10. Osborn to William Gregory, May 25, 1935. Henry Fairfield Osborn Collection, American Museum of Natural History Archives; Geoffrey Hellman *Bankers, Bones and Beetles: The First Century of the American Museum of Natural History* (Garden City, N.Y.: Natural History Press, 1969), 194; John Michael Kennedy, "Philanthropy and Science in New York City: The American Museum of Natural History, 1868–1968" (Ph.D. diss., Yale University, 1968), 208–9.

11. Elazar Barkan, *The Retreat of Scientific Racism: Changing Concepts of Race in Britain and the United States Between the World Wars* (New York: Cambridge University Press, 1992), 67.

12. Franz Boas, "The Problem of the American Negro," *Yale Review* 10 (January 1921): 384–95.

13. Ibid.

14. Ibid., 392.

15. Franz Boas Papers, American Philosophical Society, File, White, Walter: Walter White to Franz Boas, March 15, 1921.

16. Franz Boas Papers, American Philosophical Society, File, Boas Lectures: "The Races of Man," 1896.

17. Franz Boas Papers, American Philosophical Society, File, Boas Lectures: "Commencement Address at Atlanta University," May 31, 1906. See also Julia E. Liss et al., "Diasporic Identities: The Science and Politics of Race in the Work of Franz Boas and W. E. B. Du Bois, 1894–1919," *Cultural Anthropology* 13 (May 1998): 127–66.

18. Lee D. Baker, *From Savage to Negro: Anthropology and the Construction of Race, 1896–1954* (Berkeley: University of California Press, 1998), 119.

19. Harry Laughlin, *The Second International Exhibition of Eugenics Held September 22 to October 22, 1921, in Connection with the Second International Congress of Eugenics in the American Museum of Natural History, New York: An Account of the Organization of the Exhibition, the Classification of the Exhibits, the List of Exhibitors, and a Catalog and Description of the Exhibit* (Baltimore: Wilkins and Wilkins, 1923), 13, 20.

20. Ibid., 13–14.

21. Ibid., 14.

22. Ibid., 16.

23. Ibid., 21.

24. Ibid., 23.

25. Ibid., 38.

26. Ibid., 70–71.

27. Ibid., 108.

28. Kennedy, "Philanthropy and Science," 208.

29. William H. Tucker, *The Science and Politics of Racial Research* (Urbana: University of Illinois Press, 1994), 88.

30. Kennedy, "Philanthropy and Science," 208.

31. Laughlin, *Second International Exhibition*, 40.

32. Daniel J. Kevles, *In the Name of Eugenics: Genetics and the Uses of Human Heredity* (Cambridge, Mass.: Harvard University Press, 1995), 75.

33. Grant, *Passing of the Great Race*, 53.
34. Ibid., 77.
35. Ibid., 69.
36. Ibid., 78.
37. Ibid., xv.
38. Ibid., xvii.
39. Barkan, *Retreat of Scientific Racism*, 70.
40. Gregory Michael Dorr, "Segregation's Science: The American Eugenics Movement and Virginia, 1900–1980" (Ph.D. diss., University of Virginia, 2000), 250.
41. Gregory Dorr, *Segregation's Science: Eugenics and Society in Virginia* (Charlottesville: University of Virginia Press, 2008), 79.
42. Dorr, "Segregation's Science," 253.
43. Ibid., 254.
44. R. Bennett Bean, "Notes on the Body Form of Man," in *Scientific Papers of the Second International Conference of Eugenics*, vol. 2, *Eugenics in Race and State* (Baltimore: Williams and Wilkins, 1923), 17.
45. Ernest A. Hooton, "Observations and Queries as to the Effect of Race Mixture on Certain Physical Characteristics," in *Eugenics in Race and State*, 64–74.
46. W. F. Willcox, "Distribution and Increase of Negroes in the United States," in *Eugenics in Race and State*, 171.
47. Ibid., 174.
48. W. E. B. Du Bois, "Race Traits and Tendencies of the American Negro," *Annals of the American Academy of Political and Social Science* 9 (January 1897): 127–33.
49. Frederick L. Hoffman, "The Problem of Negro-White Intermixture and Intermarriage," in *Eugenics in Race and State*, 175–188.
50. Osborn, "Address of Welcome," 2.
51. Committee on Racial Problems, Joint with SSRC: Institutionalization of Infants for Controlled Data Accumulation, Letter from Knight Dunlap to Members of the Executive Committee, March 28, 1929, in National Academy of Sciences–National Research Council Archives, Division of Anthropology and Psychology Records Group.

## 4. THE NATIONAL RESEARCH COUNCIL AND THE SCIENTIFIC STUDY OF RACE

1. Committee on Racial Problems: 1928–1932, "Final Report," April 1931, in National Academy of Sciences–National Research Council Archives, Division of Anthropology and Psychology Records Group.
2. Barbara Tizard, "IQ and Race," *Nature* 247 (February 1, 1974): 316.
3. Committee on Race Characters: 1921–1922. Open letter to the Committee from Clark Wissler, Chairman, NRC Division of Anthropology and Psychology, March 24, 1921, in National Academy of Sciences–National Research Council Archives, Division of Anthropology and Psychology Records Group.

4. Committee on Racial Problems, Joint with SSRC: Institutionalization of Infants for Controlled Data Accumulation, Letter from Knight Dunlap to Members of the Executive Committee, March 28, 1929, in National Academy of Sciences–National Research Council Archives, Division of Anthropology and Psychology Records Group.

5. Daryl Michael Scott, *Contempt and Pity: Social Policy and the Image of the Damaged Black Psyche, 1880–1996* (Chapel Hill: University of North Carolina Press, 1997), xi.

6. Philip Dray, *At the Hands of Persons Unknown: The Lynching of Black America* (New York: Random House, 2002), 190; Kevin Gaines, *Uplifting the Race: Black Leadership, Politics, and Culture in the Twentieth Century* (Chapel Hill: University of North Carolina Press, 1996), 66–99.

7. August Meier, *Negro Thought in America, 1880–1915* (Ann Arbor: University of Michigan Press, 1968).

8. Kelly Miller, *Race Adjustment: Essays on the Negro in America* (New York: Neale, 1909), 44.

9. Rayford W. Logan, *The Negro in the United States: A Brief History* (Princeton: Van Nostrand, 1957).

10. W. E. B. Du Bois, *Darkwater: Voices from within the Veil* (New York: Washington Square Press, 2004), 36.

11. C. Vann Woodward, *The Strange Career of Jim Crow* (New York: Oxford University Press, 1974). See also C. Vann Woodward, *Origins of the New South: 1877–1913* (Baton Rouge: Louisiana State University Press, 1971).

12. Nicolas Lemann, *The Promised Land: The Great Black Migration and How It Changed America* (New York: Knopf, 1991).

13. Joe W. Trotter, *The Great Migration in Historical Perspective: New Dimensions of Race, Class, and Gender* (Bloomington: Indiana University Press, 1991).

14. Mark R. Schneider, *"We Return Fighting": The Civil Rights Movement in the Jazz Age* (Boston: Northeastern University Press, 2002), 20.

15. Committee on Scientific Problems of Human Migration, "Report to the Committee on Scientific Problems of Human Migration," March 8, 1923, in National Academy of Sciences–National Research Council Archives, Division of Anthropology and Psychology Records Group: appendix I, 4.

16. National Research Council, *A History of the National Research Council, 1919–1933* (Wilmington, Del., Scholarly Resources, 1974), 7.

17. National Research Council, "The National Research Council: Organization of the National Research Council," *Science* 49 (May 16, 1919): 458.

18. National Research Council, *History of the National Research Council*, 9–10.

19. Daniel J. Kevles, *In the Name of Eugenics: Genetics and the Uses of Human Heredity* (Cambridge, Mass.: Harvard University Press, 1995), 208–10; Ellen Condliffe Lagemann, *The Politics of Knowledge: The Carnegie Corporation, Philanthropy, and Public Policy* (Middletown, Conn.: Wesleyan University Press, 1989).

20. Roger L. Geiger, *To Advance Knowledge: The Growth of American Research Universities, 1900–1940* (New York: Oxford University Press, 1986), 99.

21. Ibid., 100.
22. Cole's January 8, 1914, talk at the conference was later published as Leon J. Cole, "Biological Eugenics: Relation of Philanthropy and Medicine to Race Betterment," *Journal of Heredity* 5 (1914): 305–12.
23. Ibid., 306.
24. Garland E. Allen, "The Eugenics Record Office at Cold Spring Harbor, 1910–1940: An Essay in Institutional History," *Osiris* 2 (1986): 225–64.
25. Charles B. Davenport, *Heredity of Skin Color in Negro-White Crosses* (Washington, D.C.: Carnegie Institution of Washington, 1913).
26. Allen, "Eugenics Record Office," 225–64.
27. Ibid., 5.
28. Lagemann, *Politics of Knowledge*, 4.
29. Ibid., 5–6.
30. Judith Sealander, "Curing Evils at Their Source: The Arrival of Scientific Giving," in *Charity, Philanthropy, and Civility in American History*, ed. Lawrence J. Friedman and Mark D. McGarvie (New York: Cambridge University Press, 2003), 229.
31. Merle Curti, "American Philanthropy and the National Character," *American Quarterly* 10 (winter 1958): 436.
32. Lagemann, *Politics of Knowledge*, 81.
33. Ibid., 47.
34. Ibid., 48; Geiger, *To Advance Knowledge*, 146.
35. Ibid., 147.
36. National Research Council, *History of the National Research Council*, 34–38.
37. Committee on Race Characters: 1923–1926, A. E. Jenks to Robert M. Yerkes, September 17, 1923, in National Academy of Sciences–National Research Council Archives, Division of Anthropology and Psychology Records Group; Kevles, *In the Name of Eugenics*, 80–81.
38. Daniel Kevles, "Testing the Army's Intelligence: Psychologists and the Military in World War I," *Journal of American History* 55 (December 1968): 565–81; Stephen Jay Gould, *The Mismeasure of Man* (New York: Norton, 1996), 222–63.
39. Gould, *Mismeasure of Man*, 260–62.
40. Committee on Race Characters: 1923–1926, Jenks to Yerkes, September 17, 1923.
41. Committee on Race Characters: 1923–1926, "Research Outline from the Division of Anthropology and Psychology," National Research Council, December 10, 1923, in National Academy of Sciences–National Research Council Archives, Division of Anthropology and Psychology Records Group: 2
42. Ibid., 3.
43. Committee on Race Characters: 1923–1926, Jenks to Yerkes, October 23, 1923, in National Academy of Sciences–National Research Council Archives, Division of Anthropology and Psychology Records Group.
44. Committee on Race Characters: 1923–1926, "Research Outline from the Division of Anthropology and Psychology," December 10, 1923.
45. Ibid., 6.

46. Ibid., 7–9

47. Ibid., 10.

48. Ibid., 10–11.

49. Lee D. Baker, *From Savage to Negro: Anthropology and the Construction of Race, 1896–1954* (Berkeley: University of California Press, 1998), 5, 93–94; See also Kevles, *In the Name of Eugenics*, and Nancy Stepan, *The Idea of Race in Science: Great Britain, 1800–1960* (Hamden, Conn.: Archon Books, 1982).

50. Committee on Scientific Problems of Human Migration, "Report to the Committee on Scientific Problems of Human Migration," March 8, 1923, appendix I, 1.

51. Committee on Scientific Problems of Human Migration, Clark Wissler, *Final Report of the Committee on Scientific Problems of Human Migration*, Reprint and Circular Series of the National Research Council (Washington, D.C.: National Academies Press, 1929), 7–8.

52. Committee on Scientific Problems of Human Migration, "Report to the Committee on Scientific Problems of Human Migration," March 8, 1923, 2.

53. Madison Grant, *The Passing of the Great Race: The Racial Basis of European History* (New York: Scribner, 1916); Lothrop Stoddard, *The Rising Tide of Color Against White World-Supremacy* (New York: Scribner, 1920).

54. Committee on Scientific Problems of Human Migration, "Conference on Racial Intermixture," February 17, 1923, in National Academy of Sciences–National Research Council Archives, Division of Anthropology and Psychology Records Group: 4–5, 8, 15.

55. Committee on Scientific Problems of Human Migration, "Report to the Committee on Scientific Problems of Human Migration," March 8, 1923, 2–3.

56. Committee on Scientific Problems of Human Migration, "Conference on Racial Intermixture," February 17, 1923, 7.

57. Charles B. Davenport, "The Effects of Race Intermingling," *Proceedings of the American Philosophical Society* 56 (1917): 364.

58. Committee on Scientific Problems of Human Migration, "Report to the Committee on Scientific Problems of Human Migration," March 8, 1923, appendix I, 5.

59. Ibid., 6.

60. Ibid.

61. Committee on Scientific Problems of Human Migration, "Conference on Racial Intermixture," February 17, 1923, 8.

62. Charles Davenport and Morris Steggerda, *Race Crossing in Jamaica* (Washington, D.C.: Carnegie Institution of Washington, 1929).

63. Raymond Pearl Papers, American Philosophical Society, File, National Research Council, Yerkes to Pearl, May 16, 1923.

64. Committee on Scientific Problems of Human Migration, Annual Reports, "Report and Recommendations of the Committee on Scientific Problems of Human Migrations," April 5, 1926, in National Academy of Sciences–National Research Council Archives, Division of Anthropology and Psychology Records Group: 6; Raymond Pearl, "On the

Pathological Relations Between Cancer and Tuberculosis," *Proceedings of the Society for Experimental Biology and Medicine* 26 (1928): 55–73; Raymond Pearl and Agnes Latimer Bacon, "New Data on Alcohol and Duration of Life," *Nature* 121 (1928): 15–16; and Raymond Pearl and Agnes Latimer Bacon, "Biometrical Studies in Pathology, V: The Racial and Age Incidence of Cancer and Other Malignant Tumors," *Archives of Pathology and Laboratory Medicine* 3 (1927): 963–92.

65. Edwin Black, *War Against the Weak: Eugenics and America's Campaign to Create a Master Race* (New York: Four Walls Eight Windows, 2003), 412.

66. Raymond Pearl Papers, American Philosophical Society, File, NAACP, #2, Correspondence between Walter White and Raymond Pearl.

67. Elazar Barkan, *The Retreat of Scientific Racism: Changing Concepts of Race in Britain and the United States Between the World Wars* (New York: Cambridge University Press, 1992), 212–20.

68. Raymond Pearl Papers, American Philosophical Society, File, National Research Council, Yerkes to Pearl, November 25, 1922; Davenport, "Effects of Race Intermingling," 364–68.

69. Ibid., Pearl to Yerkes, November 28, 1922.

70. Ibid., Yerkes to Pearl, November 29, 1922.

71. Quoted in William H. Tucker, *The Science and Politics of Racial Research* (Urbana: University of Illinois Press, 1994), 71.

72. Raymond Pearl and Agnes Latimer Bacon, "Biometrical Studies in Pathology, VI: The Primary Site of Cancers and Other Malignant Tumors," *Archives of Pathology and Laboratory Medicine* 6 (1928): 67–89.

73. Ibid., 81.

74. Pearl and Bacon, "Biometrical Studies in Pathology, V," 963–92.

75. Ibid., "Biometrical Studies in Pathology, VI," 80–81.

76. Transcript of the "Conference on Racial Differences," February 1928, 23, in National Academy of Sciences–National Research Council Archives, Division of Anthropology and Psychology Records Group.

77. Raymond Pearl, "Evolution and Mortality," *Quarterly Review of Biology* 3 (June 1928): 271–80.

78. Ibid., 274.

79. Committee on Scientific Problems of Human Migration, "Report to the Committee on Scientific Problems of Human Migration," March 8, 1923, appendix I, 8.

80. Wissler, *Final Report*, 15.

81. Ibid., 9, 10, 14; Committee on Scientific Problems of Human Migration, Annual Reports, "Report and Recommendations of the Committee on Scientific Problems of Human Migrations," April 5, 1926, 6.

82. Edmund Ramsden, "Social Demography and Eugenics in the Interwar United States," *Population and Development Review* 29 (December 2003): 548.

83. Wissler, *Final Report*, 17–21.

84. Ibid., 16.

## 5. COLORING RACE DIFFERENCE

1.  Committee on the Study of the American Negro, 1926–1929, "Committee on the American Negro: Proposals for the Organization of Investigations on the American Negro," n.d., in National Academy of Sciences–National Research Council Archives, Division of Anthropology and Psychology Records Group.
2.  Robert J. Terry. "The American Negro," *Science* 69 (March 29, 1929): 337–41.
3.  Committee on the Study of the American Negro, 1926–1929, R. M. Terry to G. M. Stratton, October 20, 1925.
4.  Ibid., Letter from A. E. Jenks to G. M. Stratton, January 27, 1926.
5.  Ibid., "Report of Progress, Committee on the American Negro, 1926–1927, appendix K," in National Academy of Sciences–National Research Council Archives, Division of Anthropology and Psychology Records Group.
6.  Ibid., "Committee on the American Negro: Proposals for the Organization of Investigations on the American Negro."
7.  Ibid., "Report of Progress, Committee on the American Negro, 1928–1929, appendix G."
8.  Ibid., "Report of Progress, Committee on the American Negro, 1926–1927, appendix K."
9.  Ibid., "A&P Annual Meeting: 1928."
10. Ibid., "Report of Progress, Committee on the American Negro, 1928–1929, appendix G."
11. Transcript of the "Conference on Racial Differences," February 1928, in National Academy of Sciences–National Research Council Archives, Division of Anthropology and Psychology Records Group: 3.
12. Ibid., 5
13. Ibid.
14. Fred Eggan, "Fay-Cooper Cole, 1881–1961," *American Anthropologist* 65 (June 1963): 641–48.
15. Transcript of the "Conference on Racial Differences," February 1928, 9.
16. Franz Boas, "Changes in the Bodily Form of Descendants of Immigrants," *American Anthropologist* 14 (July–September, 1912): 530–62.
17. Audrey Smedley, *Race in North America: Origin and Evolution of a Worldview* (Boulder: Westview Press, 1993), 276.
18. Ibid., 276–77.
19. Transcript of the "Conference on Racial Differences," February 1928, 16–17.
20. Ibid., 18.
21. Ibid., 19.
22. Ibid., 21.
23. Ernst Mayr, "Darwin's Impact on Modern Thought," *Proceedings of the American Philosophical Society* 139 (December 1995): 317–25.
24. Transcript of the "Conference on Racial Differences," February 1928, 19–20.
25. William B. Provine, *The Origins of Theoretical Population Genetics* (Chicago: University of Chicago Press, 2001).
26. Kenneth Ludmerer, "American Geneticists and the Eugenics Movement, 1905–1935," *Journal of the History of Biology* 2 (September 1969): 347–48, 350–51.

27. Joan Fisher Box, *R. A. Fisher: The Life of a Scientist* (New York: Wiley, 1978), 268, as quoted in Elazar Barkan, *The Retreat of Scientific Racism: Changing Concepts of Race in Britain and the United States Between the World Wars* (New York: Cambridge University Press, 1992), 222; R. A. Fisher, *The Genetic Theory of Natural Selection* (New York: Oxford University Press, 1930), 258.

28. Edward East, *Mankind at the Crossroads* (New York: Scribner's, 1923), 133.

29. Bentley Glass, "Geneticists Embattled: Their Stand Against Rampant Eugenics and Racism in America During the 1920s and 1930s," *Proceedings of the American Philosophical Society* 30 (1986): 135.

30. Transcript of the "Conference on Racial Differences," February 1928, 31.

31. Ibid., 33.

32. Melville J. Herskovits, *The American Negro: A Study in Race Crossing* (New York: Knopf, 1928), 67, 82.

33. Transcript of the "Conference on Racial Differences," February 1928, 22.

34. Ibid., 12–13.

35. Wayne C. Richard, "Joseph Peterson: Scientist and Teacher," *Peabody Journal of Education* 46 (July 1968): 3–8.

36. Joseph Peterson, *The Comparative Abilities of White and Negro Children* (Baltimore: Williams and Wilkins, 1923).

37. Joseph Peterson, "Methods of Investigating Comparative Abilities in Races," *Annals of the American Academy of Political and Social Science* 140 (November 1928): 178–85.

38. Charles H. Thompson, "The Conclusions of Scientists Relative to Racial Differences," *Journal of Negro Education* 3 (July 1934): 498.

39. Transcript of the "Conference on Racial Differences," February 1928, 36–41.

40. Ibid., 42.

41. Committee on Racial Problems, 1928–1931, Joint with SSRC, "Memo," March 26, 1928, in National Academy of Sciences–National Research Council Archives, Division of Anthropology and Psychology Records Group.

42. Transcript of the "Conference on Racial Differences," February 1928, in National Academy of Sciences–National Research Council Archives, Division of Anthropology and Psychology Records Group: 77–80.

43. Committee on Racial Problems, 1928–1931, Joint with SSRC, "Minutes of the Meeting Held January 12, 1929," January 12, 1929, in National Academy of Sciences–National Research Council Archives, Division of Anthropology and Psychology Records Group.

44. Ibid., "Final Report," April, 1931, in National Academy of Sciences–National Research Council Archives, Division of Anthropology and Psychology Records Group.

45. Ibid., Memorandum to the Members of the Conference of Directors, May 12, 1930, in National Academy of Sciences–National Research Council Archives, Division of Anthropology and Psychology Records Group.

46. Roy M. Dorcus, "Knight Dunlap: 1875–1949," *American Journal of Psychology* 63 (January 1950): 114–19.

47. Mary Ann Mason, *From Father's Property to Children's Rights: The History of Child Custody in the United States* (New York: Columbia University Press, 1994); Matthew A. Crenson, *Building the Invisible Orphanage: A Prehistory of the American Welfare System* (Cambridge, Mass.: Harvard University Press, 1998).

48. Committee on Racial Problems, 1928–1931, Joint with SSRC, Fay-Cooper Cole to Knight Dunlap, September 17, 1928, in National Academy of Sciences–National Research Council Archives, Division of Anthropology and Psychology Records Group.

49. Ibid., Franz Boas to Knight Dunlap, April 1, 1929, in National Academy of Sciences–National Research Council Archives, Division of Anthropology and Psychology Records Group.

50. Ibid., "Report Made to the SSRC Committee on Problems and Policy, August 1930, in National Academy of Sciences–National Research Council Archives, Division of Anthropology and Psychology Records Group, 1–2.

51. Ibid., 3–4.

52. Ibid., "Joint Committee on Racial Matters of the Social Science Research Council and the National Research Council," 6.

53. Ibid., "Final Report," April 31.

54. Susan Reverby, *Examining Tuskegee: The Infamous Syphilis Study and Its Legacy* (Chapel Hill: University of North Carolina Press, 2009), 24–26.

55. Charles Davenport Papers, American Philosophical Society, File, Kidder, Alfred: Davenport to Kidder, October 21, 1926.

56. Ibid., November 3, 1926.

57. Ibid., Committee on the American Negro: Davenport to Terry, January 4, 1927.

58. Ibid., November 12, 1928.

59. Ibid., March 4, 1929.

60. Ibid., File, W. P. Draper, Folder 2, 1929–1932: Davenport to Draper, July 16, 1929; W. P. Draper, Folder 1, 1923–1928: Davenport to Draper, May 25, 1928.

61. Ibid., Folder 1, 1923–1928: Draper to Davenport, March 20, 1923.

62. Ibid., February 5, 1926.

63. Ibid., Davenport to Draper, February 6, 1926.

64. Ibid., March 15, 1926.

65. Ibid., File, Todd, T. Wingate: Davenport to Todd, October 27, 1928.

66. William H. Tucker, *The Funding of Scientific Racism: Wickliffe Draper and the Pioneer Fund* (Urbana: University of Illinois Press, 2002), 30–33.

67. Charles B. Davenport, "Race Crossing in Jamaica," *Scientific Monthly* 27 (September 1928): 225–38.

68. Charles Davenport and Morris Steggerda, *Race Crossing in Jamaica* (Washington, D.C.: Carnegie Institution of Washington, 1929), 477.

69. Frank H. Hankins, "Heredity and Environment," *Social Forces* 9 (June 1931): 587–88.

70. Karl Pearson, review of *Race Crossing in Jamaica*, by Charles Davenport and Morris Steggerda, *Nature* 126 (1930): 427.

71. Glass, "Geneticists Embattled," 130–54.

72. William B. Provine, "Geneticists and the Biology of Race Crossing," *Science* 182 (November 23, 1973): 790–96.

73. William E. Castle, "Race Mixture and Physical Disharmonies," *Science* 71 (June 13, 1930): 605.

74. Ibid.

75. Charles Davenport Papers, American Philosophical Society, File, Davenport, Charles B. Lectures: "Do Races Differ in Mental Capacity, March 1924" [The date on this file must be incorrect since the lecture discusses research being carried out in Jamaica by Steggerda, research that did not begin until 1926], "Racial Traits, February 1921," and "Heredity and Race Eugenics, 1927."

76. Davenport, "Race Crossing in Jamaica."

77. Peggy Pascoe, "Miscegenation Law, Court Cases, and Ideologies of 'Race' in Twentieth-Century America," *Journal of American History* 83 (June 1996): 44–69.

78. Gregory Michael Dorr, "Segregation's Science: The American Eugenics Movement and Virginia, 1900–1980" (Ph.D. diss., University of Virginia, 2000), 515.

79. Robert R. Hurwitz, "Constitutional Law: Equal Protection of the Laws, California Anti-Miscegenation Laws Declared Unconstitutional," *California Law Review* 37 (March 1949): 127n28.

80. Tucker, *Funding of Scientific Racism*, 33–38; E. S. Cox, *White America* (Richmond, Va.: White America Society, 1923).

81. John P. Jackson Jr., *Science for Segregation: Race, Law, and the Case Against Brown v. Board of Education* (New York: New York University Press, 2005), 34–35.

## 6. BIOLOGY AND THE PROBLEM OF THE COLOR LINE

1. W. E. B. Du Bois, ed., *The Health and Physique of the Negro American* (Atlanta: Atlanta University Press, 1906).

2. Lee D. Baker, *From Savage to Negro: Anthropology and the Construction of Race, 1896–1954* (Berkeley: University of California Press, 1998), 115.

3. Ibid., 99–10, 113.

4. W. E. B. Du Bois, *The Souls of Black Folk: Essays and Sketches* (Chicago: McClurg, 1903), vii.

5. W. E. B. Du Bois, Elijah Anderson, and Isabel Eaton, *The Philadelphia Negro: A Social Study* (Philadelphia: University of Pennsylvania Press, 1996).

6. W. E. B. Du Bois, *Dusk of Dawn: An Essay Toward an Autobiography of a Race Concept* (Piscataway, N.J.: Transaction, 1983), 58.

7. Ibid., 63.

8. Wissler to Du Bois, November 31, 1905, in *The Correspondence of W. E. B. Du Bois, Volume 1: Selections, 1877–1934*, ed. Herbert Aptheker (Amherst: University of Massachusetts Press, 1977), 115.

9. Joseph Deniker, *The Races of Man: An Outline of Anthropology and Ethnology* (Freeport, N.Y.: Books for Libraries Press, 1971), 2–3.

10. Du Bois to Carnegie, May 22, 1906, in Aptheker, *Correspondence of W. E. B. Du Bois*, 121–22.

11. Franz Boas Papers, American Philosophical Society, File, Du Bois, W. E. Burghardt: Du Bois to Boas, October 11, 1905, and Boas to Du Bois, October 14, 1905.

12. Ibid., Boas Lectures: Commencement Address at Atlanta University, May 31, 1906.

13. Du Bois, *Health and Physique*, 13.

14. Ibid., 13.

15. Ibid., 16.

16. Ibid., 28.

17. Ibid., 29.

18. Ibid.

19. Ibid., 24–27.

20. Ibid., 31–58.

21. Ibid., 89–90.

22. Frederick L. Hoffman, *Race Traits and Tendencies of the American Negro* (New York: American Economic Association, 1896).

23. Ibid., 326.

24. Megan J. Wolff, "The Myth of the Actuary: Life Insurance and Frederick L. Hoffman's Race Traits and Tendencies of the American Negro," *Public Health Reports* 121 (January–February, 2006): 86.

25. Ibid., 88.

26. W. E. B. Du Bois, "Race Traits and Tendencies of the American Negro," *Annals of the American Academy of Political and Social Science* 9 (January 1897): 133.

27. Ibid., 127.

28. Du Bois, *Health and Physique*, 89–90.

29. Baker, *From Savage to Negro*. See also Elazar Barkan, *The Retreat of Scientific Racism: Changing Concepts of Race in Britain and the United States Between the World Wars* (New York: Cambridge University Press, 1992).

30. Du Bois, *Dusk of Dawn*, 100.

31. Dominic J. Capeci Jr. and Jack C. Knight, "Reckoning with Violence: W. E. B. Du Bois and the 1906 Atlanta Race Riot," *Journal of Southern History* 62 (November 1996): 740; David Levering Lewis, *W. E. B. Du Bois: Biography of a Race* (New York: Holt, 1993).

32. David L. Lewis. "Atlanta Is Swept by Raging Mob: Over 16 Negroes Reported to Be Dead," *Atlanta Constitution*, September 23, 1906, B1; David Fort Godshalk, *Veiled Visions: The 1906 Atlanta Race Riot and the Reshaping of American Race Relations* (Chapel Hill: University of North Carolina Press, 2005).

33. Capeci and Knight, "Reckoning with Violence," 749.

34. Ibid., 759.

35. Lewis, *W. E. B. Du Bois*, 3.

36. W. E. B. Du Bois, *Darkwater: Voices from within the Veil* (New York: Washington Square Press, 2004), 54–55.

37. W. E. B. Du Bois, "Purity of Blood," in *Crisis: A Record of the Darker Races, Volumes 9–10, 1914–1915* (New York: Negro Universities Press, 1969) 276; Carol M. Taylor, "W. E. B. Du Bois's Challenge to Scientific Racism," *Journal of Black Studies* 11 (1981): 449–60.

38. W. E. B. Du Bois, "Races," *Crisis* 2 (August 1911): 157.

39. David Levering Lewis, *W. E. B. Du Bois: The Fight for Equality and the American Century, 1919–1963* (New York: Holt, 2000), 235–36.

40. Ibid., 236–37.

41. Daylanne K. English, *Unnatural Selections: Eugenics in American Modernism and the Harlem Renaissance* (Chapel Hill: University of North Carolina Press, 2004), 41.

42. Lewis, *W. E. B. Du Bois: Fight for Equality*, 455–56.

43. Anthony Appiah, "The Uncompleted Argument: Du Bois and the Illusion of Race," *Critical Inquiry* 12 (autumn 1985): 4–5.

44. Jacques Barzun, *Race: A Study in Modern Superstition* (New York: Harcourt, Brace, 1937), 1.

45. Ruth Benedict, *Race: Science and Politics* (New York: Viking, 1940), v–vi.

46. Ibid., 11.

47. Ivan Hannaford, *Race: The History of an Idea in the West* (Washington, D.C.: Woodrow Wilson Center Press, 1996), 376–90.

48. Benedict, *Race*, 230–31.

49. Ibid., 233.

50. Harvard Sitkoff, *A New Deal for Blacks: The Emergence of Civil Rights as a National Issue; The Depression Decade* (New York: Oxford University Press, 1981).

51. Barzun, *Race*, 11.

52. Ibid., 7, 10.

53. Ibid., 27

54. Columbia University Rare Book and Manuscript Library, Jacques Barzun Collection, Box 6, File, Race: A Study in Superstition (vol. II): Barzun to Oliver C. Cox, April 19, 1948.

55. Rosemary Firth, review of *Race: A Study in Modern Superstition*, by Jacques Barzun, *Man* 39 (May 1939): 38–39.

56. Barkan, *Retreat of Scientific Racism*, 109.

57. Clark Wissler, review of *Race: A Study in Modern Superstition*, by Jacques Barzun, *American Historical Review* 44 (October 1938): 62.

58. Columbia University Rare Book and Manuscript Library, Jacques Barzun Collection, Box 6, File: Race: A Study in Superstition (vol. II), Barzun to Editor, *American Historical Review*, December 1, 1938.

59. Barzun, *Race*, 283.

60. Ibid., 296.

61. Madison Grant, *The Passing of the Great Race: The Racial Basis of European History* (New York: Scribner, 1916), 45.

62. William H. Tucker, *The Science and Politics of Racial Research* (Urbana: University of Illinois Press, 1994), 27.

63. Stefan Kühl, *The Nazi Connection: Eugenics, American Racism, and German National Socialism* (New York: Oxford University Press, 2002), 39.

64. Edwin Black, *War Against the Weak: Eugenics and America's Campaign to Create a Master Race* (New York: Four Walls Eight Windows, 2003), 294–95, 312–14; Kühl, *Nazi Connection*, 48–49.

## 7. RACE AND THE EVOLUTIONARY SYNTHESIS

1. Stephen Jay Gould, introduction to *Genetics and the Origin of Species*, by Theodosius Dobzhansky (New York: Columbia University Press, 1982), xxi.

2. Ernst Mayr, *The Growth of Biological Thought: Diversity, Evolution, and Inheritance* (Cambridge, Mass.: Belknap Press, 1982), 567; Jan Sapp, *Genesis: The Evolution of Biology* (New York: Oxford University Press, 2003), 143. Sapp's and Mayr's points are both based on the conclusions of Julian Huxley's classic summary of the evolutionary synthesis titled *Evolution: The Modern Synthesis* (London: Allen and Unwin, 1942).

3. Ernst Mayr, "The Role of Systematics in the Evolutionary Synthesis," in *Systematics and the Origin of Species*, ed. Ernst Mayr and William B. Provine (Cambridge, Mass.: Harvard University Press, 1980), 123–24.

4. Mayr, *Growth of Biological Thought*, 570.

5. Theodosius Dobzhansky Papers, American Philosophical Society, Restricted File: Dobzhansky to R. C. Murphy, March 7, 1947.

6. Dobzhansky, *Genetics and the Origin of Species*; Huxley, *Evolution*; George G. Simpson, *Tempo and Mode in Evolution* (New York: Columbia University Press, 1944); Mayr, *Systematics and the Origin of Species*.

7. Ernst Mayr, "Prologue: Some Thoughts on the History of the Evolutionary Synthesis," in *The Evolutionary Synthesis: Perspective on the Unification of Biology*, ed. Ernst Mayr and William B. Provine (Cambridge, Mass.: Harvard University Press, 1980), 13.

8. Ibid., 17.

9. Ernst Mayr, "Typological versus Population Thinking," in *Conceptual Issues in Evolutionary Biology*, ed. Eliot Sober (Cambridge, Mass.: MIT Press, 1994), 158; Eliot Sober, "Population Thinking, and Essentialism," *Philosophy of* Science 47 (1980): 367–68.

10. Lisa Gannett and James R. Griesemer, "The Genetics of ABO Blood Groups," in *Classical Genetic Research and Its Legacy: The Mapping of Cultures of Twentieth-Century Genetics*, ed. Hans-Jörg Rheinberger and Jean-Paul Gadillière (New York: Routledge, 2004), 161.

11. Kenneth Ludmerer, "American Geneticists and the Eugenics Movement, 1905–1935," *Journal of the History of Biology* 2 (September 1969): 337–62.

12. Elazar Barkan, *The Retreat of Scientific Racism: Changing Concepts of Race in Britain and the United States Between the World Wars* (New York: Cambridge University Press, 1992), 140.

13. Julian Huxley, "Letter to the Editor," *Eugenics Review* 29 (1938), as quoted in William B. Provine, "Geneticists and the Biology of Race Crossing," *Science* 182 (November 23, 1973): 790–96.

14. Gould, introduction, xxv.

15. Sophia Dobzhansky Coe, "Theodosius Dobzhansky: A Family Story," in *The Evolution of Theodosius Dobzhansky: Essays on His Life and Thought in Russia and America*, ed. Mark Adams, 13–28 (Princeton: Princeton University Press, 1994).

16. See chapters by Nikolai L. Kremenstov, Daniel A. Alexandrov, and Mikhail B. Konashev in Adams, *Evolution of Theodosius Dobzhansky*, 31–84.

17. Theodosius Dobzhansky, "Morgan and His School in the 1930s," in Mayr and Provine, *Evolutionary Synthesis*, 445.

18. V. B. Smocovitis, *Unifying Biology: The Evolutionary Synthesis and Evolutionary Biology* (Princeton: Princeton University Press, 1996), 133.

19. Sewell Wright, review of *Genetics and the Origin of Species*, by Theodosius Dobzhansky, *Botanical Gazette* 99 (June 1938): 955–56; Theodosius Dobzhansky Papers, American Philosophical Society, File, Sewall Wright: Wright to Dobzhansky, October 22, 1937.

20. Gould, introduction, xxvi.

21. Barkan, *Retreat of Scientific Racism*.

22. William Provine, "Genetics and Race," *American Zoologist* 26 (1986): 857–87.

23. Theodosius Dobzhansky and L. C. Dunn, *Heredity, Race, and Society* (New York: Penguin, 1947); Theodosius Dobzhansky, *The Biological Basis of Human Freedom* (New York: Columbia University Press, 1956); Bruce Wallace and Theodosius Dobzhansky, *Radiation, Genes, and Man* (New York: Holt, 1959); Theodosius Dobzhansky, *The Biology of Ultimate Concern* (New York: New American Library, 1967); Theodosius Dobzhansky, *Genetic Diversity and Human Equality* (New York: Basic Books, 1973).

24. Mayr, "Prologue."

25. Daniel A. Alexandrov, "Filipchenko and Dobzhansky: Issues in Evolutionary Genetics in the 1920s," in Adams, *Evolution of Theodosius Dobzhansky*, 49–62; William B. Provine, "Origins of the GNP Series," in *Dobzhansky's Genetics of Natural Populations: I–XLIII*, ed. R. C. Lewontin, John Moore, and William B. Provine, and Bruce Wallace, 5–76 (New York: Columbia University Press, 1981).

26. Mark B. Adams, "Towards a Synthesis: Population Concepts in Russian Evolutionary Thought, 1925–1935," *Journal of the History of Biology* 3 (spring 1970): 107–29.

27. Ernst Mayr Papers, American Philosphical Society, Transcript, May 23, 1974, Session II (B) Afternoon, 7.

28. Sewell Wright Papers, American Philosophical Society, File, 1943, Dobzhansky, Theodosius: Dobzhansky to Wright, September 28, 1943.

29. Theodosius Dobzhansky, "Geographical Variation in Lady-Beetles," *American Naturalist* 67 (March–April 1933): 98–99.

30. Provine, "Origins of the GNP," 59.

31. Mayr, *Systematics and the Origin of Species*.

32. Provine, "Origins of the GNP," 11.

33. Leslie C. Dunn Papers, American Philosophical Society, File, Provine, William: "Reply from Dobzhansky." Although there is no date on Dobzhansky's letter to Provine, the original Provine solicitation was sent on August 5, 1971. All the responses to Provine from other scientists were dated in the second half of 1971.

34. Theodosius Dobzhansky Papers, American Philosophical Society, Notebooks, Box 1, 1953: Dobzhansky to Dunn, January 23, 1954 (letter taped into journal).

35. Audrey Smedley, *Race in North America: Origin and Evolution of a Worldview* (Boulder: Westview Press, 1993), 303–10.

36. Ralph Bunche, "What Is Race?" in *Ralph Bunche: Selected Speeches and Writings*, ed. Charles P. Henry, 207–20 (Ann Arbor: University of Michigan Press, 1995). Originally published in Ralph Bunche, *A World View of Race* (Washington, D.C.: Associates in Negro Folk Education, 1936).

37. Ibid., 214–15.

38. Ibid., 219.

39. Ibid., 207.

40. Dobzhansky, *Genetics and the Origin of Species*, 47.

41. Ibid., 60–62. Emphasis added.

42. Ibid., 60, 62–63.

43. Ibid., 62–63.

44. Theodosius Dobzhansky, "The Race Concept in Biology," *Scientific Monthly* 52 (February 1941): 161.

45. Columbia University Rare Book and Manuscript Library, Jacques Barzun Collection, Box 6, File, Race: A Study in Superstition (vol. II): Dobzhansky to Barzun, December 18, 1942.

46. Dobzhansky, "Race Concept," 161–62.

47. Ibid., 163–64.

48. Mayr and Provine, *Evolutionary Synthesis*, 29.

49. Dobzhansky, "Race Concept," 165.

50. Ashley Montagu Papers, American Philosophical Society, Box 12, Dobzhansky to Montagu, May 22, 1944.

51. Leslie C. Dunn Papers, American Philosophical Society, File, Dobzhansky, Theodosius—Dunn Correspondence #5, 1946–7: Dobzhansky to Dunn, March 10, 1947.

52. Mary F. Lyon, "L. C. Dunn and Mouse Genetic Mapping," in *Perspectives on Genetics: Anecdotal, Historical, and Critical Commentaries, 1987–1998*, ed. James F. Crow and William F. Dove, 161–66 (Madison: University of Wisconsin Press, 2000); Melinda Gormley, "Geneticist L. C. Dunn: Politics, Activism, and Community" (Ph.D. diss., Oregon State University, 2006), 112–22.

53. Joe Cain, "Co-opting Colleagues: Appropriating Dobzhansky's 1936 Lectures at Columbia," *Journal of the History of Biology* 35 (summer 2002): 207–19.

54. Leslie C. Dunn Papers, American Philosophical Society, File, Dobzhansky, Theodosius—Dunn Correspondence #1, 1936–7: Dobzhansky to Dunn, May 27, 1936.

55. Ibid., Dunn to Dobzhansky, May 4, 1937.

56. Theodosius Dobzhansky, "Leslie Clarence Dunn, 1893–1974," in *A Biographical Memoir* (Washington, D.C.: National Academy of Sciences, 1978). For more detailed information on Dunn's life and work, see Gormley, "Geneticist L. C. Dunn"; and Dorothea Bennett, "L. C. Dunn and His Contribution to T-Locus Genetics," *Annual Review of Genetics* 11 (1977): 1–12.

57. L. C. Dunn, "A Biological View of Race Mixture," *Proceedings of the American Sociological Society* 19 (1925): 47.

58. L. C. Dunn, "An Anthropometric Study of Hawaiians of Pure and Mixed Blood," *Papers of the Peabody Museum of American Archaeology and Ethnology* 11 (1928): 91–211.

59. Gormley, "Geneticist L. C. Dunn," 98–101.

60. Charles Davenport Papers, American Philosophical Society, L. C. Dunn Correspondence: Dunn to Merriam, July 3, 1935.

61. Garland E. Allen, "The Eugenics Record Office at Cold Spring Harbor, 1910–1940: An Institutional History," *Osiris* 2 (1986): 225–64.

62. Dobzhansky, "Leslie Clarence Dunn," 85.

63. Gormley, "Geneticist L. C. Dunn," 491–96.

64. Bennett, "L. C. Dunn," 1.

65. L. C. Dunn and Theodosius Dobzhansky, *Heredity, Race, and Society* (New York: Penguin, 1946), 6.

66. Ibid., 5.

67. Ibid., 10.

68. Robert Cook and Jay L. Lush, "Genetics for the Millions: An Unfinished Story," *Journal of Heredity* 38 (October 1947): 299–305.

69. Dunn and Dobzhansky, *Heredity, Race, and Society*, 91–92.

70. Ibid., 108–9.

71. Ibid., 108.

72. Ibid., 113.

73. Bentley Glass, review of *Heredity, Race, and Society*, by Theodosius Dobzhansky and L. C. Dunn, *Quarterly Review of Biology* 22 (June 1947): 152.

74. Cook and Lush, "Genetics for the Millions," 303–4.

75. Ashley Montagu Papers, American Philosophical Society, Box, Dobzhansky: Dobzhansky to Montagu, March 9, 1948.

76. Ibid., May 1, 1948.

77. Leslie C. Dunn Papers, American Philosophical Society, File, Dobzhansky, Theodosius—Dunn Correspondence #5, 1946–7: Dobzhansky to Dunn, November 25, 1946.

78. See, for example, Carl L. Hubbs, "Concepts of Homology and Analogy," *American Naturalist* 78 (July–August, 1944): 289–307; Roger Lewin, "When Does Homology Mean Something Else?" *Science* 237 (September 1987): 1570; N. Jardine, "The Concept

of Homology in Biology," *British Journal for the Philosophy of Science* 18 (1967): 125–39; and G. P. Wagner, "The Biological Homology Concept," *Annual Review of Ecology and Systematics* 20 (1989): 51–69.

79. Jonathan Marks, *What It Means to be 98% Chimpanzee: Apes, People, and Their Genes* (Berkeley: University of California Press, 2002), 73–74.

80. Susan Sperling, "Ashley Montagu (1905–1999)," *American Anthropologist* 102 (September 2000): 583–88; Michelle Brattain, "Race, Racism, and Antiracism: UNESCO and the Politics of Presenting Science to the Postwar Public," *American Historical Review* 112 (December 2007): 1393–94.

81. Ashley Montagu Papers, American Philosophical Society, File, Lieberman, Leonard: Lieberman to Montagu (with Montagu's answers), August 22, 1994.

82. M. F. Ashley Montagu, "The Genetical Theory of Race, and Anthropological Method," *American Anthropologist* 44 (July–September 1942): 370, 373.

83. Ibid., 375–76.

84. Ibid., 369.

85. Brattain, "Race, Racism, and Antiracism," 1395.

86. Ashley Montagu Papers, Lieberman to Montagu, August 22, 1994.

87. Andrew P. Lyons, "The Neotenic Career of M. F. Ashley Montagu," in *Race and Other Misadventures: Essays in Honor of Ashley Montagu In His Ninetieth Year*, ed. Larry T. Reynolds and Leonard Lieberman (Dix Hills, N.Y.: General Hall, 1996), 10–11.

88. Ashley Montagu Papers, American Philosophical Society, File, Dobzhansky, Theodosius: Montagu to Dobzhansky, May 23, 1944.

89. Ashley Montagu, *Man's Most Dangerous Myth: The Fallacy of Race* (New York: Columbia University Press, 1945), 244.

90. Ibid., 45.

91. Frank H. Hankins, review of *Man's Most Dangerous Myth: The Fallacy of Race*, by Ashley Montagu, *Annals of the American Academy of Political and Social Science* 227 (May 1943): 191–92.

92. Clyde Kluckhohn, review of *Man's Most Dangerous Myth: The Fallacy of Race*, by Ashley Montagu, *Isis* 34 (summer 1943): 419.

93. Bentley Glass, review of *Man's Most Dangerous Myth: The Fallacy of Race*, by Ashley Montagu, *Quarterly Review of Biology* 21 (March 1946): 128.

94. Aldous Huxley, foreword to Montagu, *Man's Most Dangerous Myth*, vii.

95. See, for example, M. F. Ashley Montagu, "The African Origins of the American Negro and His Ethnic Composition," *Scientific Monthly* 58 (January 1944): 58–65; M. F. Ashley Montagu, "Physical Characteristics of the American Negro," *Scientific Monthly* 59 (July 1944): 56–62; M. F. Ashley Montagu, "Intelligence of Northern Negroes and Southern Whites in the First World War," *American Journal of Psychology* 58 (April 1945): 161–88; M. F. Ashley Montagu, "Blood Group Factors and Ethnic Relationships," *Science* 103 (March 1, 1946): 284; and Ashley Montagu and Benjamin Pasamanick, "Racial Intelligence," *Scientific Monthly* 66 (January 1948): 81–82.

## 8. CONSOLIDATING THE RACE CONCEPT IN BIOLOGY

1. James T. Patterson, *Brown v. Board of Education: A Civil Rights Milestone and Its Troubled Legacy* (New York: Oxford University Press, 2001), 67; Michael J. Klarman, *Unfinished Business: Racial Equality in American History* (New York: Oxford University Press, 2007), 148.

2. Aldon D. Morris, "A Retrospective on the Civil Rights Movement: Political and Intellectual Landmarks," *Annual Review of Sociology* 25 (1999): 517–39; Harvard Sitkoff, *New Deal for Blacks: The Emergence of Civil Rights as a National Issue* (New York: Oxford University Press, 1978); John Dittmer, *Local People: The Struggle for Civil Rights in Mississippi* (Urbana: University of Illinois Press, 1994); Charles W. Eagles, "Toward New Histories of the Civil Rights Era," *Journal of Southern History* 66 (November 2000), 815–48; Richard M. Dalfiume, "The 'Forgotten Years' of the Negro Revolution," *Journal of American History* 55 (June 1968): 90–106.

3. James T. Patterson, *Grand Expectations: The United States, 1945–1974* (New York: Oxford University Press, 1996), 389–95.

4. C. Vann Woodward, *The Strange Career of Jim Crow* (New York: Oxford University Press, 2002), 146–47.

5. *Argument: The Oral Argument Before the Supreme Court in Brown v. Board of Education of Topeka, 1952–55*, ed. Leon Friedman (New York: Chelsea House, 1969), 330.

6. Ibid., 330.

7. See, for example, Walter A. Jackson, *Gunnar Myrdal and America's Conscience: Social Engineering and Racial Liberalism, 1938–1987*, Fred W. Morrison Series in Southern Studies (Chapel Hill: University of North Carolina Press, 1990); Oliver C. Cox, "The Modern Caste School of Race Relations," *Social Forces* 21 (December 1942): 218–26.

8. Gunnar Myrdal, *An American Dilemma: The Negro Problem and American Democracy* (New Brunswick, N.J.: Transaction, 1996). Regarding the Clarks' studies, see, for example, Herbert Garfinkel, "Social Science Evidence and the School Segregation Cases," *Journal of Politics* 21 (February 1958): 37–59; William H. Tucker, *The Science and Politics of Racial Research* (Urbana: University of Illinois Press, 1994), 138–53; Richard Kluger, *Simple Justice: The History of Brown v. Board of Education and Black America's Struggle for Equality* (New York: Knopf, 1980), 315–45; Patterson, *Brown v. Board of Education*, 67–68.

9. Sanjay Mody, "*Brown* Footnote Eleven in Historical Context: Social Science and the Supreme Court's Quest for Legitimacy," *Stanford Law Review* 54 (April 2002): 795, 804, 806, 808, 814–29.

10. Myrdal, *American Dilemma*; David W. Southern, "An American Dilemma After Fifty Years: Putting the Myrdal Study and Black-White Relations in Perspective," *History Teacher* 28 (February 1995): 227–53.

11. Myrdal, *American Dilemma*, 89.

12. Ibid., 15.

13. Kluger, *Simple Justice*, 582–616; Patterson, *Brown v. Board of Education*, 46–69.

14. Frances Gaither, "Democracy: The Negro's Hope," *New York Times*, April 2, 1944, BR7.

15. W. E. B. Du Bois, review of *An American Dilemma*, by Gunnar Myrdal, *Phylon* 5 (2nd quarter 1944): 114–24.

16. Oscar Handlin, review of *An American Dilemma*, by Gunnar Myrdal, *New York Times Book Review*, April 21, 1963, 1, as quoted in Jackson, *Gunnar Myrdal and America's Conscience*, 294.

17. Ibid., 330–31.

18. John P. Jackson Jr., *Science for Segregation: Race, Law, and the Case Against Brown v. Board of Education* (New York: New York University Press, 2005), 41.

19. Ibid., 91.

20. Curt Stern, "Model Estimates of the Frequency of White and Non-White Segregants in the American Negro," *Acta Genetica Basel* 4 (1953): 281–98.

21. James V. Neel, "Curt Stern, 1902–1981," *Annual Review of Genetics* 17 (1983): 1–10; James V. Neel, "The William Allan Memorial Award," *American Journal of Human Genetics* 27 (1975): 135–39.

22. D. Peter Snustad and Michael J. Simmons, *Principles of Genetics* (New York: Wiley, 2003), 773.

23. Neel, "William Allan Memorial Award," 136–37.

24. Neel, "Curt Stern," 7.

25. Curt Stern Papers, American Philosophical Society, File, Lecture: "Why Do People Differ?": Stern to NAACP, Rochester, February 11, 1946.

26. Ibid., File, Stern, C.: "The Biology of the Negro": Leon Svirsky to Stern, August 20, 1954.

27. Curt Stern, "The Biology of the Negro," *Scientific American* 191 (October 1954): 81.

28. Ibid., 85.

29. Curt Stern Papers, Correspondence: undated and unsigned correspondence to Stern.

30. Ibid., "Crackpot Letters": C. L Barnett to Stern, August 13, 1955; Mrs. John Lansdell Howerton to Stern, n.d.

31. Ibid., "The Biology of the Negro," Correspondence: Marcus Julius Frogstein to Stern, October 11, 1954.

32. Ibid., H. J. Romm to Stern, September 29, 1954; Arndt to Stern, October 8, 1954.

33. Stern, "Biology of the Negro," 82.

34. Ashley Montagu, *Statement on Race: An Annotated Elaboration and Exposition of the Four Statements on Race Issued by the United Nations Educational, Scientific, and Cultural Organization* (New York: Oxford University Press, 1972), 1.

35. Michelle Brattain, "Race, Racism, and Antiracism: UNESCO and the Politics of Presenting Science to the Postwar Public," *American Historical Review* 112 (December 2007): 1386–1413.

36. International Eugenics Congress, *Problems in Eugenics: Papers Communicated to the First International Eugenics Congress* (London: Eugenics Education Society, 1912); Edwin Black, *War Against the Weak: Eugenics and America's Campaign to Create a Master Race* (New York: Four Walls Eight Windows, 2003), 236, 245

37. Leslie C. Dunn Papers, American Philosophical Society, Series I, File, Boas, Franz: Boas to Dunn, October 25, 1937.

38. Ashley Montagu Papers, American Philosophical Society, Box 30, File, Lieberman, Leonard: undated interview of Montagu by Leonard Lieberman, Andrew Lyons, and Harriet Lyons. This interview was published by the authors in *Current Anthropology* 36 (December 1995): 841.

39. Montagu, *Statement on Race*, 4–6.

40. Ibid., 7–9.

41. Ibid., 9.

42. Ibid., 9–13.

43. Ashley Montagu, "No Scientific Basis for Race Bias Found by World Panel of Experts," *New York Times*, July 18, 1950, 1.

44. Ashley Montagu, "The Myth of Race," *New York Times*, July 19, 1950, 30.

45. Ashley Montagu, "Let's Forget About the Myth of Race," *Hartford Courant*, July 20, 1950, 8.

46. Ashley Montagu Papers, American Philosophical Society, Box 7, File, Castle, W. E.: Castle to Montagu, April 21, 1951.

47. Ibid., Box 11, File, Dunn, L. C.: Dunn to Montagu, March 3, 1950.

48. Ibid., Box 12, File, Dobzhansky, T.: Dobzhansky to Montagu, October 16, 1950.

49. Ibid., Montagu to Dobzhansky, May 23, 1944.

50. Ibid., Dobzhansky to Montagu, January 26, 1951.

51. Ashley Montagu, "In Memoriam: Osman Hill," *Journal of Anatomy* 120 (1975): 387–90.

52. Osman Hill, letter to the editor, *Man* 51 (January 1951): 16–17.

53. Ashley Montagu Papers, Box 12, File, Dobzhansky, T.: Dobzhansky to Montagu, January 26, 1951.

54. "UNESCO and Race," *Man* 51 (May 1951): 64.

55. Ashley Montagu Papers, American Philosophical Society, Box 12, File, Dobzhansky, T.: Dobzhansky to Montagu, February 24, 1951.

56. Jackson, *Science for Segregation*, 61; Brattain, "Race, Racism, and Antiracism," 1393–94.

57. Montagu, *Statement on Race*, 137–38.

58. Leslie C. Dunn Papers, American Philosophical Society, File, UNESCO, 1951: Alfred Métraux to Dunn, June 26, 1951.

59. Montagu, *Statement on Race*, 139.

60. Ibid., 143.

61. Ibid., 140.

62. Ibid., 10, 145.

63. Leslie C. Dunn Papers, File, UNESCO, 1951: Dunn to Métraux, February 26, 1952; Métraux to Dunn, April 18, 1952.

64. William Provine, "Genetics and Race," *American Zoologist* 26 (1986): 877; UNESCO, *The Race Concept: Results of an Inquiry* (Paris: UNESCO, 1952).

65. Leslie C. Dunn Papers, File, UNESCO, 1951: R. A. Fisher to Métraux, October 3, 1951.

66. Ibid., S. E. Washburn to Métraux, October 15, 1951.

67. Ibid., File, Provine, William: Dobzhansky to Provine, undated; Wright to Provine, August 17, 1971; Dunn to Provine, November 17, 1971.

68. Ibid., Darlington to Provine, August 13, 1971; Stern to Provine, August, 18, 1971.

69. Elazar Barkan, *The Retreat of Scientific Racism: Changing Concepts of Race in Britain and the United States Between the World Wars* (New York: Cambridge University Press, 1992), 341–42.

70. Elazar Barkan, "The Politics of the Science of Race: Ashley Montagu and UNESCO's Anti-racist Declarations," in *Race and Other Misadventures: Essays in Honor of Ashley Montagu in His Ninetieth Year*, ed. Larry T. Reynolds and Leonard Lieberman (Dix Hills, N.Y.: General Hall, 1996), 103–4.

71. Ibid., 103.

72. Provine, "Genetics and Race," 877.

73. Theodosius Dobzhansky Papers, American Philosophical Society, Notebooks Box 1: "June 14 Paris–Iceland," 1951; Ashley Montagu Papers, Box 12, File, Dobzhansky, T.: Dobzhansky to Montagu, February 24, 1951.

74. Brattain, "Race, Racism, and Antiracism," 1413.

75. Ibid., 1386–87, 1407–12.

76. Ibid., 1413.

77. Leslie C. Dunn Papers, File, UNESCO, 1951: R. A. Fisher to Métraux, October 3, 1951; Provine, "Genetics and Race," 875.

78. UNESCO, *Race Concept*, 40–60.

79. L. C. Dunn, *Race and Biology* (Paris: UNESCO, 1965), 7.

80. UNESCO, *Race Concept*, 79–80.

81. Montagu, *Statement on Race*, 61, 65.

82. See, for example, Lee D. Baker, *From Savage to Negro: Anthropology and the Construction of Race, 1896–1954* (Berkeley: University of California Press, 1998); and Tucker, *Science and Politics*.

83. C. P. Snow, *The Two Cultures* (New York: Cambridge University Press, 1998), 3, 4.

84. V. B. Smocovitis, *Unifying Biology: The Evolutionary Synthesis and Evolutionary Biology* (Princeton: Princeton University Press, 1996), 152–53, 167.

85. The Reminiscences of Leslie C. Dunn (1960), p. 886, in the Columbia Center for Oral History.

86. Ronald G. Walters, "Uncertainty, Science, and Reform in Twentieth-Century America," in *Scientific Authority and Twentieth-Century America*, ed. Ronald G. Walters (Baltimore: Johns Hopkins University Press, 1997), 6.

87. Ibid., 8–10.

88. Paul Boyer, *By the Bomb's Early Light: American Thought and Culture at the Dawn of the Atomic Age* (New York: Pantheon, 1985), 354; Margot A. Henriksen, *Dr. Strangelove's America: Society and Culture in the Atomic Age* (Berkeley: University of California Press, 1997), 58.

89. Henriksen, *Dr. Strangelove's America*, 54–56.

90. Michael Polanyi, "Scientific Outlook: Its Sickness and Cure," *Science* 125 (March 15, 1957): 480.

91. Steve Fuller, "Being There with Thomas Kuhn: A Parable for Postmodern Times," *History and Theory* 31 (October 1992): 261.

92. Thomas Borstelmann, *The Cold War and the Color Line: American Race Relations in the Global Arena* (Cambridge, Mass.: Harvard University Press, 2001), 55.

93. Borstelmann, *Cold War*, 54–55; Mary L. Dudziak, *Cold War Civil Rights: Race and the Image of American Democracy* (Princeton: Princeton University Press, 2000), 18–24.

94. Morris, "Retrospective on the Civil Rights Movement," 518.

95. Dudziak, *Cold War Civil Rights*, 13.

96. Borstelmann, *Cold War*, 59–61.

97. Dudziak, *Cold War Civil Rights*, 25–26, 82.

98. Harry S. Truman, "Special Message to the Congress on Civil Rights," February 2, 1948, *Public Papers of the Presidents of the United States: Harry S. Truman, 1948* (Washington, D.C.: Government Printing Office, 1964): 121–26, as quoted in Dudziak, *Cold War Civil Rights*, 82.

99. Brian Urquhart, *Ralph Bunche: An American Life* (New York: Norton, 1993), 230–32.

100. Ben Keppel, *The Work of Democracy: Ralph Bunche, Kenneth B. Clark, Lorraine Hansberry, and the Cultural Politics of Race* (Cambridge, Mass.: Harvard University Press, 1995), 64.

101. Charles S. Johnson, "American Minorities and Civil Rights in 1950," *Journal of Negro Education* 20 (summer 1951): 489.

## 9. CHALLENGES TO THE RACE CONCEPT

1. Carleton Coon, *The Origin of Races* (New York: Knopf, 1962).

2. Arthur Jensen, "How Much Can We Boost IQ and Scholastic Achievement?" *Harvard Educational Review* 39 (spring 1969): 1–123.

3. Richard E. Green et al., "A Draft Sequence of the Neanderthal Genome," *Science* 310 (2010): 721; see also Alan R. Templeton, "Out of Africa Again and Again," *Nature* 416 (2002): 45–51; Svante Pääbo, "The Mosaic That Is Our Genome," *Nature* 421 (2003): 409–12.

4. Milford Wolpoff and Rachel Caspari, *Race and Human Evolution* (New York: Simon and Schuster, 1997).

5. Coon, *Origin of Races*, 656.

6. Ibid., vii, 482–587.

7. Ibid., 662.

8. Ibid., 661.

9. Ernst Mayr, review of *The Origin of Races*, by Carleton Coon, *Science* 138 (October 19, 1962): 420–22.

10. Frederick S. Hulse, review of *The Origin of Races*, by Carleton Coon, *American Anthropologist* 65 (June 1963): 685–87.

11. Malcolm F. Farmer, "Stepping Stone Toward an Understanding of Man's Development," *Phylon* 24 (2nd quarter 1963): 203.

12. Barbara Tuchman, "Reviewers' Choice, 1962," *Chicago Daily Tribune*, December 2, 1962, E8. Emphasis added.

13. Theodosius Dobzhansky Papers, American Philosophical Society, File, Ernst Mayr: Dobzhansky to Coon, October 23, 1962; John P. Jackson Jr., *Science for Segregation: Race, Law, and the Case Against Brown v. Board of Education* (New York: New York University Press, 2005), 162–170.

14. Michael Lerner Papers, American Philosophical Society, File, Dobzhansky, Theodosius #5: Dobzhansky to Lerner (address to colleague), December 17, 1962; Margaret Mead, "Scientist Reviewers Beware," *Science* 141 (July 26, 1964): 312–13.

15. Theodosius Dobzhansky, "Possibility That Homo Sapiens Evolved Independently 5 Times Is Vanishingly Small," *Current Anthropology* 4 (October 1963): 360–66.

16. Jackson, *Science for Segregation*, 187.

17. Carleton S. Coon, "Comments," *Current Anthropology* 4 (October 1963): 366.

18. Jackson, *Science for Segregation*, 99–103, 189.

19. Theodosius Dobzhansky Papers, File, Mayr, Ernst 1962: Mayr to Dobzhansky, November 1, 1962.

20. Ibid., Dobzhansky to Simpson, Mayr, and Strauss, November 9, 1962.

21. Theodosius Dobzhansky, *Mankind Evolving: The Evolution of the Human Species* (New Haven: Yale University Press, 1962), 253.

22. Ibid., 262–66.

23. Ibid., 178.

24. Ibid., 182.

25. Ibid., 183.

26. Ibid., 185.

27. Ashley Montagu, "What Is Remarkable About Varieties of Man Is Likeness, Not Differences," *Current Anthropology* 4 (October 1963): 362.

28. Frank B. Livingstone and Theodosius Dobzhansky, "On the Non-Existence of Human Races," *Current Anthropology* 3 (1962): 280.

29. Frank B. Livingstone. "Anthropological Implications of Sickle Cell Gene Distribution in West Africa," *American Anthropologist* 60 (1958) 533–62.

30. Livingstone and Dobzhansky, "On the Non-Existence of Human Races."

31. Ibid.

32. Jensen, "How Much Can We Boost IQ," 3

33. Ibid., 117.

34. Ibid., 29.

35. See, for example, Thomas F. Jackson, *From Civil Rights to Human Rights: Martin Luther King, Jr., and the Struggle for Economic Justice* (Philadelphia: University of Pennsylvania

Press, 2006); Wesley C. Hogan, *Many Minds, One Heart: SNCC's Dream for a New America* (Chapel Hill: University of North Carolina Press, 2007); Michael K. Honey, *Going Down Jericho Road: The Memphis Strike, Martin Luther King's Last Campaign* (New York: Norton, 2007).

36. Harold M. Schmeck Jr., "Nobel Winner Urges Research on Racial Heredity," *New York Times*, October 18, 1966, 9; "Possible Metallurgical and Astronomical Approaches to the Problem of Environment versus Ethnic Heredity," in National Academy of Sciences–National Research Council Archives, Division of A Psychology Records Group.

37. File: Comments, Institutions: General, Stanford University News Service Press Release, October 17, 1966, in National Academy of Sciences–National Research Council Archives, Division of Anthropology and Psychology Records Group.

38. Press Release, Michigan State University, October 20, 1967, in National Academy of Sciences–National Research Council Archives, Central File: Committee on Science and Public Policy, Study on Gene Pool Deterioration: Proposed.

39. Shockley to Gardner, October 13, 1967, in National Academy of Sciences–National Research Council Archives, Central File: Committee on Science and Public Policy: Study on Gene Pool Deterioration: Proposed; Lee to Shockley, November 17, 1967.

40. Margaret Mead, "Introductory Remarks," in *Science and the Concept of Race*, ed. Margaret Mead, Theodosius Dobzhansky, Ethel Tobach, and Robert E. Light, (New York: Columbia University Press, 1969): 3.

41. Ashley Montagu Papers, American Philosophical Society, Dobzhansky Box: Dobzhansky to Montagu, January 26, 1951.

## 10. NATURALIZING RACISM: THE CONTROVERSY OVER SOCIOBIOLOGY

1. E. B. Ford, "Theodosius Grigorievich Dobzhansky: 26 January 1900–18 December 1975," *Biographical Memoirs of Fellows of the Royal Society* 23 (November 1977): 62.

2. Francisco J. Ayala, "Theodosius Dobzhansky, 1900–1975," in *A Biographical Memoir* (Washington, D.C.: National Academy of Sciences, 1985), 166.

3. Richard Lewontin Papers, American Philosophical Society, File, Dobzhansky, Professor Theodosius, #IV: typed article of an obituary of Dobzhansky that would be published in the *Egyptian Journal of Genetics and Cytology*.

4. Theodosius Dobzhansky, "Leslie Clarence Dunn: November 2, 1893–March 19, 1974," in *A Biographical Memoir* (Washington, D.C.: National Academy of Sciences), 86.

5. Theodosius Dobzhansky, "Leslie Dunn, Well-Known Geneticist," *Washington Post*, March 23, 1974, D5.

6. Edward O. Wilson, *Sociobiology: The New Synthesis* (Cambridge, Mass.: Belknap Press, 1975).

7. William M. Dugger, "Sociobiology for Social Scientists: A Critical Introduction to E. O. Wilson's Evolutionary Paradigm," *Social Science Quarterly* 62 (June 1981): 229; Richard Lewontin, interview by author, November 2, 1995, Cambridge, Mass.

8. V. B. Smocovitis, "Unifying Biology: The Evolutionary Synthesis and Evolutionary Biology," *Journal of the History of Biology* 25 (March 1992): 1.

9. Wilson, *Sociobiology: The New Synthesis*, 4.

10. Edward O. Wilson, "What Is Sociobiology?" *Society* (September/October 1978): 10.

11. Wilson, *Sociobiology: The New Synthesis*, 5.

12. Edward O. Wilson, *On Human Nature* (Cambridge, Mass.: Harvard University Press, 1978), 43. Wilson, *Sociobiology: The New Synthesis*, 585.

13. R. A. Fisher, *The Genetical Theory of Natural Selection* (Oxford: Clarendon Press, 1930), 256–65; Julian Huxley, *Evolution: The Modern Synthesis* (London: Harper, 1943), 572–78.

14. Desmond Morris, *The Naked Ape: A Zoologist's Study of the Human Animal* (New York: McGraw-Hill, 1967); Robert Ardry, *African Genesis: A Personal Investigation into the Animal Origins and Nature of Man* (New York: Atheneum, 1961); Konrad Lorenz, *Studies in Animal and Human Behavior* (Cambridge, Mass.: Harvard University Press, 1971); Richard Dawkins, *The Selfish Gene* (New York: Oxford University Press, 1976); David Barash, *The Whisperings Within* (New York: Harper and Row, 1979).

15. "A Genetic Defense of the Free Market," *Business Week*, April 10, 1978, 100; Maya Pines, "Is Sociobiology All Wet?" *Psychology Today* 11 (June 1978): 24.

16. Marshall Sahlins, "The Use and Abuse of Biology," in *The Sociobiology Debate: Readings on Ethical and Scientific Issues*, ed. Arthur Caplan (New York: Harper and Row, 1978), 424–27.

17. John Pfeiffer, review of *Sociobiology*, by Edward O. Wilson, *New York Times Book Review*, July 27, 1975, 15–16.

18. Mary Jane West-Eberhard, "Born: Sociobiology," *Quarterly Review of Biology* 51 (March 1976): 92.

19. David P. Barash, "Ethology, Ecology, and Evolution: Getting It Together," *Ecology* 57 (March 1976): 399–400.

20. E. O. Wilson Papers, Library of Congress, Series 21765, Box 1, File, Material on Sociobiology/Letters on Sociobiology: Lorenz to Wilson, August 19, 1975; Darlington to Wilson, May 26, 1975.

21. Stephen Jay Gould, "Biological Potential vs. Biological Determinism," in *Ever Since Darwin: Reflections in Natural History*, 251–59 (New York: Norton, 1977).

22. Henry Louis Gates Jr., "Critical Remarks," in *Anatomy of Racism*, ed. David Theo Goldberg (Minneapolis: University of Minnesota Press, 1990), 326.

23. Howard Winant and Michael Omi, *Racial Formation in the United States: From the 1960s to the 1980s* (New York: Routledge, 1986), 110.

24. Ibid., 110.

25. Bruce J. Schulman, *The Seventies: The Great Shift in American Culture, Society, and Politics* (New York: Da Capo Press, 2002), 58–84; Michael Klarman, *Unfinished Business: Racial Equality in American History* (New York: Oxford University Press, 2007), 183–198; Thomas Segrue, *The Origins of the Urban Crisis: Race and Inequality in Post-War Detroit* (Princeton: Princeton University Press, 1996).

26. Dorothy Nelkin and Susan M. Lindee, *The DNA Mystique: The Gene as a Cultural Icon* (New York: Freeman, 1995), 2, 194.

27. Troy Duster, *Backdoor to Eugenics* (New York: Routledge, 1990), 93.

28. Wilson, *Sociobiology: The New Synthesis*, 550; Richard Lewontin, "The Apportionment of Human Diversity," *Evolutionary Biology* 6 (1972): 381–98.

29. Robert Lunbeck, "Anti-Racism Group Attacks Wilson's 'Sociobiology,' " *Harvard Crimson*, December 3, 1975.

30. Vernon Reynolds, "Sociobiology and Race Relations," in *The Sociobiology of Ethnocentrism: Evolutionary Dimensions of Xenophobia, Discrimination, Racism, and Nationalism*, ed. Vernon Reynolds, Vincent Falger, and Ian Vine, 208–15 (Athens: University of Georgia Press, 1986), 212.

31. Daniel J. Kevles, *In the Name of Eugenics: Genetics and the Uses of Human Heredity* (Cambridge, Mass.: Harvard University Press, 1995), 83; Arthur Jensen, "How Much Can We Boost IQ and Scholastic Achievement," *Harvard Educational* Review 39 (spring 1969): 1–123; Arthur Jensen, "Race and the Genetics of Intelligence: A Reply to Lewontin," *Bulletin of the Atomic Scientists* 26 (May 1970): 17–23.

32. Wilson, "What Is Sociobiology?" 47–48.

33. Pierre van den Berghe, "Race and Ethnicity: A Sociobiological Perspective," *Ethnic and Racial Studies* 1 (October 1978): 403.

34. Ibid, 402; Edward O. Wilson, *Sociobiology: The Abridged Edition* (Cambridge, Mass.: Belknap Press, 1980), 314, 315; Reynolds, Falger, and Vine, *Sociobiology of Ethnocentrism*.

35. Wilson, *Sociobiology: The New Synthesis*, 286–87, 564–65.

36. Pierre van den Berghe, *Race and Racism: A Comparative Perspective* (New York: Wiley, 1967), 18.

37. Van den Berghe, "Race and Ethnicity," 404.

38. David Barash, *The Hare and the Tortoise* (New York: Viking, 1986), 144.

39. David Barash, *The Whisperings Within* (New York: Harper and Row, 1979), 154, 232.

40. J. Philippe Rushton, "Comments," *Social Science and Medicine* 31 (1990): 905–10; J. Philippe Rushton, "Genetic Similarity Theory: Intelligence and Human Mate Choice," *Ethology and Sociobiology* 9 (1988): 45–57; J. Philippe Rushton, "Evidence for Genetic Similarity Detection in Human Marriage," *Ethology and Sociobiology* 6 (1985): 183–87.

41. J. Philippe Rushton, *Race, Evolution, and Behavior* (New Brunswick, N.J.: Transaction, 1995), 113–46; Adolph Reed Jr. "Intellectual Brown Shirts," in *The Bell Curve Debate: History, Documents, and Opinions*, ed. Russell Jacoby and Naomi Glauberman (New York: New York Times Books, 1995), 268.

42. Richard J. Herrnstein and Charles A. Murray, *The Bell Curve: Intelligence and Class Structure in American Life* (New York: Free Press, 1994), 642–43.

43. Adam Miller, "Professors of Hate," *Rolling Stone*, October 20, 1994; William H. Tucker, *The Science and Politics of Racial Research* (Urbana: University of Illinois Press, 1994), 291–92.

44. Vernon Reynolds, Vincent Falger, and Ian Vine, "Introduction by the Editors," and Robin I. M. Dunbar, "Sociobiological Explanations and the Evolution of Ethnocentrism," in Reynolds, Falger, and Vine, *Sociobiology of Ethnocentrism*, xv–xx; 48–59.

45. E. O. Wilson Papers, Library of Congress, Series 20196, Box 6, File, Wilson, E. O. Defense of Sociobiology: Wilson to Frank M. Carpenter, December 9, 1975.

46. Richard Lewontin Papers, File, Wilson, E. O.: Lewontin to Wilson, October 28, 1975.

47. Edward O. Wilson, "What Is Sociobiology?" *Society* 15 (September–October 1978): 10.

48. Edward O. Wilson, *Naturalist* (Washington, D.C.: Island Press, 1994), 336.

49. Elizabeth Allen, Barbara Beckwith, Jon Beckwith, Steven Chorover, and David Culver, et al., "Against 'Sociobiology,'" *New York Review of Books*, November 13, 1975.

50. Richard Lewontin Papers, File, Wilson, E. O.: Wilson to Robert B. Silvers, November 10, 1975.

51. Ibid., Wilson to Lewontin, December 17, 1975.

52. E. O. Wilson to Gould, November 10, 1975, Stephen Jay Gould Papers, Box 525, Correspondence, Incoming, M–Z, 1975–1979, M1437. Department of Special Collections, Stanford University Libraries, Stanford, Calif.

53. Garland Allen to Gould, February 23, 1977, Stephen Jay Gould Papers, Box 524, Correspondence, Incoming A–L, 1975–1979, M1437. Department of Special Collections, Stanford University Libraries, Stanford, Calif.

54. Edward. O. Wilson, "Academic Vigilantism and the Political Significance of Sociobiology," *BioScience* 26 (March 1976): 183, 187–90.

55. "Report to Eastern Regional SftP Conference, April 15–17, 1977 at Voluntown," Stephen Jay Gould Papers, Box 607, M1437. Department of Special Collections, Stanford University Libraries, Stanford, Calif.

56. Sociobiology Study Group, "Sociobiology: Tool for Social Oppression," *Science for the People* 8 (March 1976): 7; Robin Marantz Henig, "10 Years Later . . . Science for the People: Revolution's Evolution," *BioScience* 29 (June 1979): 341–44.

57. Robin Marantz Henig, "Burning Darwin to Save Marx," *Harpers*, December 1978, 31.

58. Sociobiology Study Group, "Sociobiology: Another Biological Determinism," *BioScience* 26 (March 1976): 280.

59. Edward O. Wilson, "Human Decency Is Animal," *New York Times Magazine*, October 12, 1975, 50; Barbara Chasin, "Sociobiology: A Sexist Synthesis," *Science for the People* 9 (May–June 1977): 30.

60. Sociobiology Study Group, "Sociobiology: A New Biological Determinism," in Caplan, *Sociobiology Debate*, 280.

61. Minutes of May 10, June 7, and November 8, 1977 meetings, Stephen Jay Gould Papers, Box 607, M1437. Department of Special Collections, Stanford University Libraries, Stanford, Calif.

62. Henig, "10 Years Later"; Richard C. Lewontin, "Science for the People," *BioScience* 29 (September 1979): 509.

63. Ullica Segerstråle, *Defenders of the Truth: The Battle for Science in the Sociobiology Debate and Beyond* (New York: Oxford University Press, 2000), 23.

64. Richard Lewontin Papers, File, Wilson, E. O.: Wilson to Silvers, November 19, 1975.

65. E. O. Wilson Papers, Library of Congress, Box 11, File, Le Monde: response to RL Interview, October 1980, Wilson to Editors at Le Monde, October 21, 1980.

66. Wilson, "What Is Sociobiology?" 191.

67. Neil Jumonville, "The Cultural Politics of the Sociobiology Debate," *Journal of the History of Biology* 35 (2002): 569–93.

68. Ibid., 191.

69. Charles J. Lumsden and Edward O. Wilson, *Promethean Fire: Reflections on the Origin of Mind* (Cambridge, Mass.: Harvard University Press, 1983), 43.

70. E. O. Wilson Papers, Library of Congress, Series 20196, Box 6, File, Political Uses of Theory by New Right: Richard Lynn to Wilson, July 7, 1976.

71. Ibid., Wilmot Robertson to Wilson, August, 29, 1977.

72. Richard Lewontin Papers, Box 5, File: E. O. Wilson, Lewontin to Wilson, July 19, 1979.

73. Ibid.

74. Stephen Jay Gould Papers, Box 607, Minutes of June 26, 1979 meeting, Sociobiology Study Group, Science for the People. Department of Special Collections, Stanford University Libraries, Stanford, Calif.

75. E. O. Wilson Papers, Library of Congress, Series 20196, Box 6, File: Political Uses of Theory by New Right, undated note.

76. Henig, "10 Years Later," 341.

77. Richard Lewontin Papers, File, Herrnstein, Professor R. J.: Herrnstein to Lewontin, June 29, 1973; Lewontin to Herrnstein, November 20, 1973.

78. Ibid., File, Shockley, William: Lewontin to Shockley, October 19, 1973.

79. Richard C. Lewontin and Jack L. Hubby, "A Molecular Approach to the Study of Genic Heterozygosity in Natural Populations. I. The Number of Alleles at Different Loci in *Drosophila pseudoobscura*," *Genetics* 54 (1966): 546–95; Richard C. Lewontin and Jack L. Hubby, "A Molecular Approach to the Study of Genic Heterozygosity in Natural Populations. II. Amount of Variation and Degree of Heterozygosity in Natural Populations of *Drosophila pseudoobscura*," *Genetics* 54 (1966): 595–609.

80. Jeffrey Powell, review of *The Genetic Basis of Evolutionary Change*, by Richard Lewontin, *BioScience* 25 (February, 1975): 118.

81. Lewontin, "Apportionment of Human Diversity," 397

82. Ibid., 396.

83. Maryellen Ruvolo and Mark Seielstad, "The Apportionment of Human Diversity 25 Years Later," in *Thinking About Evolution: Historical, Philosophical, and Political Perspectives*, ed. Rama S. Singh, Costas B. Krimbas, Diane B. Paul, and John Beatty, 141–51 (New York: Cambridge University Press, 2001).

84. Wilson, *Sociobiology: The New Synthesis*, 550.

85. Guido Barbujani, Arianna Magagni, Eric Minch, L. Luca Cavalli-Sforza, "An Apportionment of Human DNA Diversity," *Proceedings of the National Academy of Sciences* 94 (1997): 4518.

86. Richard Lewontin, *The Genetic Basis of Evolutionary Change* (New York: Columbia University Press, 1974).

87. Marcus W. Feldman, review of *The Genetic Basis of Evolutionary Change*, by Richard Lewontin, *Quarterly Review of Biology* 50 (September 1975), 293.

88. Powell, review of *Genetic Basis*, 118.

89. Michael Ruse, review of *The Genetic Basis of Evolutionary Change*, by Richard Lewontin, *Philosophy of Science* 43 (June 1976): 303.

90. Segerstråle, *Defenders of the Truth*, 36.

91. Ibid., 37; Wilson, *Sociobiology: The New Synthesis*, 70.

92. Edward O. Wilson, "In the Queendom of Ants: A Brief Autobiography," in *Studying Animal Behavior: Autobiographies of the Founders*, edited by Donald A. Dewsbury (Chicago: University of Chicago Press, 1989), 481.

## 11. RACE IN THE GENOMIC AGE

1. Oswald T. Avery, Colin M. MacLeod, and Maclyn McCarty, "Studies on the Chemical Nature of the Substance Inducing Transformation of Pneumococcal Types," *Journal of Experimental Medicine* 79 (1944): 137–58.

2. Erwin Chargaff, "Preface to a Grammar of Biology: A Hundred Years of Nucleic Acid Research," *Science* 172 (1971): 637–42

3. Brenda Maddox, *Rosalind Franklin: The Dark Lady of DNA* (New York: HarperCollins, 2002).

4. James Watson and Francis Crick, "Molecular Structure of Nucleic Acids: A Structure for Deoxyribose Nucleic Acid," *Nature* 171 (1953): 737–38.

5. Michael R. Dietrich, "Paradox and Persuasion: Negotiating the Place of Molecular Evolution within Evolutionary Biology," *Journal of the History of Biology* 31 (1998): 85–111. See also Joel B. Hagen, "Naturalists, Molecular Biologists, and the Challenges of Molecular Evolution," *Journal of the History of Biology* 32 (1999): 321–41; Ernst Mayr, *The Growth of Biological Thought: Diversity, Evolution, and Inheritance* (Cambridge, Mass.: Belknap Press, 1982); Theodosius Dobzhansky, "Biology, Molecular and Organismic," *American Zoologist* 4 (1964): 218–37.

6. Richard C. Lewontin and Jack L. Hubby, "A Molecular Approach to the Study of Genic Heterozygosity in Natural Populations. I. The Number of Alleles at Different Loci in *Drosophila pseudoobscura*," *Genetics* 54 (1966): 546–95; Richard C. Lewontin and Jack L. Hubby, "A Molecular Approach to the Study of Genic Heterozygosity in Natural Populations. II. Amount of Variation and Degree of Heterozygosity in Natural Populations of *Drosophila pseudoobscura*," *Genetics* 54 (1966): 595–609; Motoo Kimura, "Evolutionary Rate at the Molecular Level," *Nature* 217 (1968): 624–26; Willi Hennig, *Phylogenetic Systematics* (Urbana: University of Illinois Press, 1966).

7. Michel Morange, *A History of Molecular Biology* (Cambridge, Mass.: Harvard University Press, 1998), 249.

8. Committee on Mapping and Sequencing the Human Genome, National Research Council, *Mapping and Sequencing the Human Genome* (Washington, D.C.: National Academies Press, 1988), 5–6.

9.  Daniel J. Kevles and Leroy Hood, *Code of Codes: Scientific and Social Issues in the Human Genome Project* (Cambridge, Mass.: Harvard University Press, 1992).

10. Jenny Reardon, *Race to the Finish: Identity and Governance in the Age of Genomics* (Princeton: Princeton University Press, 2005); J. S. Alper and J. Beckwith, "Is Racism a Central Problem for the Human Genome Diversity Project?" *Politics and Life Science* 18 (1999): 285–88.

11. Eric. T. Juengst, "The Human Genome Project and Bioethics," *Kennedy Institute of Ethics Journal* 1 (1991): 71–72.

12. Raja Mishra, "The Quest to Map the Human Genome Ends with a Truce," *Boston Globe*, June 27, 2000, C5.

13. Rick Weiss and Justin Gillis, "Teams Finish Mapping Human DNA," *Washington Post*, June 27, 2000, A1.

14. F. S. Collins and M. K. Mansoura, "The Human Genome Project: Revealing the Shared Inheritance of All Humankind," *Cancer* 92 (2001): S221.

15. G. Barbujani, A. Magagni, E. Minch, L. L. Cavalli-Sforza, "An Apportionment of Human DNA Diversity," *Proceedings of the National Academy of Sciences* 94 (1997): 4516–19; David Serre and Svante Pääbo, "Evidence for Gradients of Human Genetic Diversity Within and Among Continents," *Genome Research* 14 (2004): 1679–85; J. P. A. Ioannidis, E. E. Ntzani, T. A. Trikalinos, " 'Racial' Differences in Genetic Effects for Complex Diseases," *Nature Genetics* 36 (2004): 1312–18; M. W. Foster and R. R. Sharp, "Race, Ethnicity, and Genomics: Social Classifications as Proxies of Biological Heterogeneity," *Genome Research* 12 (2002): 844–50.

16. L. L. Cavalli-Sforza, P. Menozzi, and A. Piazza, *The History and Geography of Human Genes* (Princeton: Princeton University Press, 1993); M. Feldman, R. C. Lewontin, M. C. King, "A Genetic Melting Pot," *Nature* 424 (2003): 374; Svante Pääbo, "The Mosaic That Is Our Genome," *Nature* 421 (2003): 409–12.

17. William Stanton, *The Leopard's Spots: Scientific Attitudes Toward Race in America, 1815–59* (Chicago: University of Chicago Press, 1960); George M. Fredrickson, *Racism: A Short History* (Princeton: Princeton University Press, 2002); Audrey Smedley, *Race in North America: Origin and Evolution of a Worldview* (San Francisco: Westview Press, 1993); Stephen Jay Gould, *The Mismeasure of Man* (New York: Norton, 1996); Ashley Montagu, *Man's Most Dangerous Myth: The Fallacy of Race* (Walnut Creek, Calif.: AltaMira Press, 1997); Daniel J. Kevles, *In the Name of Eugenics: Genetics and the Uses of Human Heredity* (Cambridge, Mass.: Harvard University Press, 1995).

18. J. L. Mountain, N. Risch, "Assessing Genetic Contributions to Phenotypic Differences Among 'Racial' and 'Ethnic' Groups," *Nature Genetics* 36 (2004): S48–S53; D. A. Hinds et al. "Whole-Genome Patterns of Common DNA Variation in Three Human Populations," *Science* 307 (2005): 1072–79.

19. Reanne Frank, "What to Make of It? The (Re)emergence of a Biological Conceptualization of Race in Health Disparities Research," *Social Science and Medicine* 64 (2007) 1977–83.

20. N. Risch, E. Burchard, E. Ziv, H. Tang, "Categorization of Humans in Biomedical Research: Genes, Race, and Disease," *Genome Biology* 3 (2002): 2007.1–2007.12.

21. Robin M. Henig, "The Genome in Black and White (and Gray)," *New York Times Magazine*, October 10, 2004, 47.

22. Morris W. Foster, "Looking for Race in All the Wrong Places: Analyzing the Lack of Productivity in the Ongoing Debate About Race and Genetics," *Human Genetics* 126 (2009): 355–62.

23. L. Luca Cavalli-Sforza, "The Human Genome Diversity Project: Past, Present, and Future," *Nature Reviews Genetics* 6 (April 2005): 333–40.

24. Reardon, *Race to the Finish*, 4–6, 92–97.

25. Ibid., 92.

26. Michael Dodson and Robert Williamson, "Indigenous Peoples and the Morality of the Human Genome Diversity Project," *Journal of Medical Ethics* 25 (1999): 205.

27. Reardon, *Race to the Finish*, 159–60.

28. The International HapMap Consortium. "The International HapMap Project," *Nature* 426 (2003): 789–96.

29. Jennifer A. Hamilton, "Revitalizing Difference in the HapMap: Race and Contemporary Human Genetic Variation Research," *Journal of Law, Medicine, and Ethics* 36 (2008): 471–77.

30. International HapMap Project, How Are Ethical Concerns Being Addressed, 2012, http://hapmap.ncbi.nlm.nih.gov/ethicalconcerns.html.en.

31. International HapMap Project, Guidelines for Referring to the HapMap Populations in Publications and Presentations, 2012, http://hapmap.ncbi.nlm.nih.gov/citinghapmap.html.

32. Hamilton, Revitalizing Difference in the HapMap," 474.

33. P. C, Ng, Q. Zhao, S. Levy, R. L. Strausberg, and J. C. Venter, "Individual Genomes Instead of Race for Personalized Medicine," *Clinical Pharmacology and Therapeutics* 84 (2008): 306–9.

34. David Jones, "How Personalized Medicine Became Genetic, and Racial: Werner Kalow and the Formations of Pharmacogenetics," *Journal of the History of Medicine and Allied Sciences* 68 (2011): 1–48; H. Kim, R. Kim, A. J. Wood, C. M. Stein, "Molecular Basis of Ethnic Differences in Drug Disposition and Response," *Annual Review of Pharmacology and Toxicology* 41 (2001): 815, 850.

35. A. Bress, S. R. Patel, M. A. Perera, et al., "Effect of NQo1 and CYP4F2 Genotypes on Warfarin Dose Requirements in Hispanic-Americans and African-Americans," *Pharmacogenomics* 13 (2012), 1925–35; D. Si, J. Wang, Y. Zhang, et al., "Distribution of CYP2C9*13 Allele in the Chinese Han and the Long-Range Haplotype Containing CYP2C9*13 and CYP2C19*2," *Biopharmaceuticals and Drug Disposition* 33 (2012), 342–45; F. H. Hatta, A. Helldén, K. E. Hellgren, et al., "Search for the Molecular Basis of Ultra-Rapid CYP2C9-Catalysed Metabolism: Relationship Between SNP IVS8-109A⊠T and the Losartan Metabolism Phenotype in Swedes," *European Journal of Clinical Pharmacology* 68 (2012), 1033–42.

36. Ng, Zhao, Levy, Strausberg, and Venter, "Individual Genomes," 307–8.

37. http://www.genome.gov/Pages/About/OD/OPG/DesigningGeneticists/RCooper-Health_Disparities.pdf.

38. Understanding the Role of Genomics in Health Disparities: Toward a New Research Agenda. National Institutes of Health Meeting, September 24–26, 2008, University of Maryland.

39. Theodosius Dobzhansky, *Mankind Evolving: The Evolution of the Human Species* (New Haven: Yale University Press, 1962), 253.
40. Roberta D. Baer, Erika Arteaga, Karen Dyer, et al., "Concepts of Race and Ethnicity Among Health Researchers: Patterns and Implications," *Ethnicity and Health* 18 (2013): 211–25; Targeted Planned Enrollment Table, 2012, http://grants.nih.gov/grants/funding/women_min/guidelines_amended_10_2001.htm.
41. National Institutes of Health, 2012, http://grants.nih.gov/grants/funding/424/SF424R-R_enrollment.doc.
42. Timothy R. Rebbeck and Pamela Sankar, "Ethnicity, Ancestry, and Race in Molecular Epidemiologic Research," *Cancer Epidemiology Biomarkers and Prevention* 14 (2005): 2467–71.

## EPILOGUE: DOBZHANSKY'S PARADOX AND THE FUTURE OF RACIAL RESEARCH

1. Mindy Thompson Fullilove, "Abandoning 'Race' as a Variable in Public Health Research: An Idea Whose Time Has Come," *American Journal of Public Health* 88 (September 1998): 1297–98.
2. W. E. B. Du Bois, ed., *The Health and Physique of the Negro American* (Atlanta: Atlanta University Press, 1906).
3. Ashley Montagu, *Man's Most Dangerous Myth: The Fallacy of Race* (Walnut Creek, Calif.: AltaMira Press, 1997).
4. Theodosius Dobzhansky, *Mankind Evolving: The Evolution of the Human Species* (New Haven: Yale University Press, 1962), 253.
5. Richard Lewontin, "The Apportionment of Human Diversity," *Evolutionary Biology* 6 (1972): 397.
6. Esteban González Burchard, Elad Ziv, Natasha Coyle, et al., "The Importance of Race and Ethnic Background in Biomedical Research and Clinical Practice," *New England Journal of Medicine* 348 (2003): 1170–75.
7. Nancy Krieger et al., "Race/Ethnicity, Gender, and Monitoring Socioeconomic Gradients in Health: A Comparison of Area-Based Socioeconomic Measures; The Public Health Disparities Geocoding Project," *American Journal of Public Health* 93 (2003): 1655–71.
8. Ichiro Kawachi, Norman Daniels, and Dean E. Robinson, "Health Disparities by Race and Class: Why Both Matter," *Health Affairs* 24 (2005): 343–52.
9. R. Dawn Comstock, Edward M. Castillo, and Suzanne P. Lindsay, "Four-Year Review of the Use of Race and Ethnicity in Epidemiologic and Public Health Research," *American Journal of Epidemiology* 159 (2004): 619.
10. Camara Phyllis Jones, "Invited Commentary: 'Race,' Racism, and the Practice of Epidemiology," *American Journal of Epidemiology* 154 (2001): 299–304; Camara Phyllis Jones, "Levels of Racism: A Theoretical Framework and a Gardener's Tale," *American Journal of Public Health* 90 (2000): 1212–15.
11. Noah A. Rosenberg et al., "Genetic Structure of Human Populations," *Science* 298 (2002): 2381–85.

12. Sahotra Sarkar, *Genetics and Reductionism* (New York: Cambridge University Press, 1998), 71.
13. S. S. Lee, "Pharmacogenomics and the Challenge of Health Disparities," *Public Health Genomics* 12 (2009): 170.
14. Theodosius Dobzhansky, *Genetics and the Origin of Species* (New York: Columbia University Press, 1982), 61. Emphasis added.
15. AncestrybyDNA, 2012, http://www.ancestrybydna.com/ancestry-dna-testing-options.php.
16. AncestryDNA, 2012, http://dna.ancestry.com/#experienceAncestry.
17. Deborah A. Bolnick, Duana Fullwiley, Troy Duster, et al., "The Science and Business of Genetic Ancestry Testing," *Science* 318 (October 19, 2007): 400.
18. Keith Wailoo, Alondra Nelson, and Catherine Lee, "Genetic Claims and the Unsettled Past," in *Genetics and the Unsettled Past: The Collision of DNA, Race, and History*, ed. Keith Wailoo, Alondra Nelson, and Catherine Lee (New Brunswick, N.J.: Rutgers University Press, 2012), 2.
19. William Provine, "Genetics and Race," *American Zoologist* 26 (1986): 857–87.
20. Du Bois, *Health and Physique of the Negro American*; P. C, Ng, Q. Zhao, S. Levy, R. L. Strausberg, and J. C. Venter, "Individual Genomes Instead of Race for Personalized Medicine," *Clinical Pharmacology and Therapeutics* 84 (2008): 306–9.
21. Theodosius Dobzhansky, *Mankind Evolving: The Evolution of the Human Species* (New Haven: Yale University Press, 1962), 266–67.
22. Ashley Montagu, *Statement on Race: An Annotated Elaboration and Exposition of the Four Statements on Race Issued by the United Nations Educational, Scientific, and Cultural Organization* (New York: Oxford University Press, 1972), 65.
23. David Levering Lewis, *W. E. B. Du Bois: The Fight for Equality and the American Century, 1919–1963* (New York: Holt, 2000), 473.
24. W. E. B. Du Bois, *Dusk of Dawn: An Essay Toward an Autobiography of a Race Concept* (Piscataway, N.J.: Transaction, 1983), 326.

# BIBLIOGRAPHY

## ARCHIVAL SOURCES CONSULTED

American Museum of Natural History, New York
    Henry F. Osborn Papers
American Philosophical Society, Philadelphia
    American Eugenics Society Records
    Franz Boas Papers
    Charles B. Davenport Papers
    Theodosius Dobzhansky Papers
    Leslie C. Dunn Papers
    Eugenics Record Office Papers
    Genetics Society of America Papers
    Michael Lerner Papers
    Richard Lewontin Papers
    Ashley Montagu Papers
    Thomas Hunt Morgan Papers
    Raymond Pearl Papers
    Herbert Spencer Jennings Papers
    Curt Stern Papers
    Sewall Wright Papers
Columbia Center for Oral History Collection, New York
    Reminiscences of Theodosius Dobzhansky
    Reminiscences of Leslie C. Dunn
Columbia University Rare Book and Manuscript Library, New York

Jacques Barzun Papers
Library of Congress, Washington, D.C.
E.O. Wilson Papers
National Academy of Sciences Archives, Washington, D.C.
 National Research Council Collection
Rockefeller Foundation Archives, Tarrytown, N.Y.
 Social Science Research Council Collection
 Laura Spelman Rockefeller Collection
Stanford University Archives, Palo Alto, Calif.
 Stephen Jay Gould Papers
University of Massachusetts, Amherst, Mass.
 W. E. B. Du Bois Papers

## A NOTE ON SOURCES

It would be an oversimplification to say that all books begin with a single ancestral source. But in the case of this work, that would largely be true. In the mid-1990s, I read a review essay by Paula Fass in *Reviews in American History* titled "Of Genes and Men" (vol. 20 [June 1992]: 235–41). That essay, critical of Carl Degler's book *In Search of Human Nature: The Decline and Revival of Darwinism in American Social Thought* (New York: Oxford University Press, 1991), helped me, as a young doctoral student, begin to shape my thoughts about the relationship between biology and society and the often uncritical way that biological explanations for a wide range of social phenomena were quickly being embraced by natural and social scientists, and even by those (Degler was her primary target) in the humanities. Fass pointed to two important issues (among many) in Degler's work that I reacted to. First, the claim by *In Search of Human Nature* that "racism is largely irrelevant to sociobiology"—chapter 10 of the present volume was written in rebuttal to Degler's wrongheaded analysis of sociobiology and race. Second, Fass attacked Degler's assertion that "facts—not ideology—can govern belief." It was this second point that interested me most and led me on a path to interrogate the meanings of race over the course of the twentieth century, particularly how science was considered an objective arbiter of the truth about what race was and what it was not.

My approach to thinking and writing about the biological sciences has been informed by an interdisciplinary background in history, public health, and biology. I have been particularly influenced by my time as a researcher in the Molecular Laboratories at the American Museum of Natural History (AMNH). Working at the museum shaped considerably how, as a historian, I think about the nature of scientific practice. As a nonscientist with complete access to and participation in a molecular laboratory, I was always a bit of a fish out of water. But as evolutionary theorists, the scientists at the AMNH were historically minded, and in that way our professional objectives overlapped: we all sought to develop an understanding of our present by reconstructing the past.

The nature and practice of science have faced considerable scrutiny by philosophers, historians, and scientists, among others. At the center of this discussion, as the philosopher of science Michael Ruse put its, is "whether science should be considered something different and special—something with independent standards which in some way guarantees its truth and importance," or whether science is "basically just a product of the same general culture as most everything else, no worse but certainly no better than those who produce it." (*Mystery of Mysteries: Is Evolution a Social Construction?* [Cambridge, Mass.: Harvard University Press, 1999], 9). At the extremes of this debate are the ideas of Karl Popper and Thomas Kuhn. Whereas Popper's hypothetico-deductive method maintains that science is both testable and falsifiable, Kuhn's belief in scientific revolutions claims that all scientific knowledge and practice are relative to the scientific paradigm—the authoritative research program that dominates science between revolutions. This does not mean that science is somehow unreal, but it does mean that it is not necessarily possible to falsify scientific theories within their paradigms (Thomas H. Kuhn, *The Structure of Scientific Revolutions* [Chicago: University of Chicago Press, 1962]; Karl R. Popper, *The Logic of Scientific Discovery* [New York: Harper and Row, 1968]). What my own experiences as a historian with training in the natural sciences has shown me is that on the one hand, science operates within a specific cultural and historical milieu. On the other hand, however, over time the scientific method tests data, theories, and ideas and discards science that does not cut the proverbial mustard. In the end, science can be both a social construction and an objective search for truth.

I visited approximately ten archives in writing this book and looked at twenty-five manuscript collections in that process. The collections at the American Philosophical Society (APS) in Philadelphia provided the most fertile material for this study, and the correspondence and other materials in its collections provided great insight into the history of the biological race concept. Theodosius Dobzhansky's papers were, as can be seen from his prominence in this book, at the center of my research, and by the time I was done working my way through them, I felt a closeness to my subject and a deep sadness as I read through the final pages of his journal written just days before he died of cancer in 1975. I would encourage scholars to mine these journals carefully; they hold wonderful material about Dobzhansky's experiences traveling the world—from the Brazilian rain forest to the Yosemite Valley—collecting his specimens and meeting many people along the way. Much of the journal is written in Russian, despite a written promise in the 1940s to begin writing only in English so that his daughter could someday share all his reflections. He did not keep that promise, and the journals, spanning almost fifty years (from his days in Russia to his death) constantly switch back and forth between Russian and English. Despite what was once a good grasp of the Russian language, I can't claim to have digested much of what was written in Russian, and another scholar is sure to find rich material in those entries.

For the history of eugenics and race, collections at the APS, the AMNH, and the National Academy of Sciences Archives were incredibly useful. The Stephen Jay Gould Papers at Stanford and the E. O. Wilson Papers at the Library of Congress were unprocessed when I utilized them. As these collections are processed, I suspect that more information on race and sociobiology will become available. Finally, the W. E. B. Du Bois Papers at the University of Massachusetts, Amherst, Archives have been digitized since my visit there, and I suspect that in time more information will be revealed about Du Bois's thoughts on the creation of his project on the *Health and Physique of the Negro American*.

As with all projects, I owe much of my thinking in the area of race and science to the important works that preceded my own. The following areas of historiography have been the most influential to this work, and the books and articles cited below are not an exhaustive accounting of works in the field or, for that matter, works cited in this book. Rather, I describe works that had a significant impact on this book, focusing primarily on books instead of articles. The notes are also an accounting of the state of the field.

## Pre-Twentieth-Century Race and Science

The best surveys of pre-twentieth-century racial science in the United States are Bruce R. Dain, *A Hideous Monster of the Mind: American Race Theory in the Early Republic* (Cambridge, Mass.: Harvard University Press, 2002); William R. Stanton, *The Leopard's Spots: Scientific Attitudes Toward Race in America, 1815–1859* (Chicago: University of Chicago Press, 1960); and Ann Fabian, *The Skull Collectors: Race, Science, and America's Unburied Dead* (Chicago: University of Chicago Press, 2010). See also Stephen Jay Gould, *The Mismeasure of Man* (New York: Norton, 1996); George Fredrickson, *The Black Image in the White Mind* (Middletown, Conn.: Wesleyan University Press, 1971); and, Drew Gilpen Faust, *The Ideology of Slavery: Proslavery Thought in the Antebellum South, 1830–1860* (Baton Rouge: Louisiana State University Press, 1981); Thomas Jefferson, *Notes on the State of Virginia* (Chapel Hill: University of North Carolina Press, 1955), 138–40, 143. For a broader discussion of Jefferson's role in the formation of a distinctly American conception of racial science, see Dain, *Hideous Monster of the Mind*; Alexander O. Boulton, "The American Paradox: Jeffersonian Equality and Racial Science," *American Quarterly* 47, no. 3 (1995): 467–92; Paul Finkelman, *Slavery and the Founders: Race and Liberty in the Age of Jefferson* (Armonk, N.Y.: M.E. Sharpe, 1996); Nell Irvin Painter, *The History of White People* (New York: Norton, 2011).

## Race, Genetics, and Eugenics

Historians who have explored the history of genetics and eugenics in the context of racial science have generally approached the topic as institutional histories examining internal developments in the field. For example, Daniel Kevles's seminal work on eugenics, *In the Name of Eugenics: Genetics and the Uses of Human Heredity* (Cambridge, Mass.: Harvard University Press, 1995), is a history of the relationship between eugenic and genetic research programs and examines the internal changes in these disciplines that helped to create the nature and texture of modern racial science. Similarly, Mike

Hawkins's *Social Darwinism in European and American Thought, 1860–1945: Nature as Model and Nature as Threat* (New York: Cambridge University Press, 1997) tackles the issue of racial science through the lens of social Darwinism. Nancy Stepan's *The Idea of Race in Science: Great Britain, 1800–1960* (Hamden, Conn.: Archon Books, 1982) has race and science at its core. Stepan suggests that after World War II the study of human diversity superseded the study of race in the sciences. The ascension of population genetics and the downfall of typological thinking facilitated this change. Stepan's book focuses primarily on science and race in Great Britain. The story of the evolution of race and science, while parallel in some ways between the United States and Britain before World War II, has a different trajectory in postwar America. While studies of human groups do begin to shift away from typological to population studies, racial science remains a powerful element of postwar biological thought in the United States.

William B. Provine's seminal articles on race, biology, and eugenics ("Geneticists and the Biology of Race Crossing," *Science* 182 [November 23, 1973]: 790–96, and "Genetics and Race," *American Zoologist* 26 [1986]: 857–87) offer an important perspective in the earlier literature on race and genetics. Unlike some of the more institutionally focused histories that examined genetics and race, Provine's periodization of how geneticists conceptualized race differences continues to provide insight into the theories and actions of these formative thinkers in the field.

The most recent foray into this subject is Mark A. Largent, *Breeding Contempt: The History of Coerced Sterilization in the United States* (New Brunswick, N.J.: Rutgers University Press, 2011), a narrative that broadens the context of forced sterilization into a history that begins in the mid-nineteenth century and continues to the turn of the twenty-first. In Largent's telling, eugenics is only part of the story behind coerced sterilization in the United States, a procedure begun in the nineteenth century by American physicians to prevent crime and punish criminals. There is, however, barely any mention of the role that race played in coerced sterilizations, either in the context of black-white or white-white ethnic differences.

Finally, Alondra Nelson's important book *Body and Soul: The Black Panther Party and the Fight Against Medical Discrimination* (Minneapolis: University of Minnesota Press, 2009) has a short but important discussion (45–47) of W. E. B. Du Bois's seminal work on race, health, and science, *The Health and Physique of the Negro American*. In *Body and*

*Soul* Nelson shows how Du Bois understood how "the arbitrariness of the American racial categories" had a "significant bearing on both the corporeal and social well-being of African Americans" and that the race concept was "explicitly linked to health and medicine." While the present volume argues that Du Bois's work in this area anticipated twentieth-century critiques of the race concept, *Body and Soul* argues that it was similarly influential "in future health activist projects, including those of the Black Panther Party," in framing arguments about "the quantity and quality of medical facilities for African Americans" and the attention to "racial health disparities as a key cause of concern," among others. See also Elof Axel Carlson, *The Unfit: A History of a Bad Idea* (Cold Spring Harbor, N.Y.: Cold Spring Harbor Laboratory Press, 2001); Mark Pittenger, *American Socialists and Evolutionary Thought, 1870–1920*, History of American Thought and Culture (Madison: University of Wisconsin Press, 1993); Michael G. Kenny, "Toward a Racial Abyss: Eugenics, Wickliffe Draper, and the Origins of the Pioneer Fund," *Journal of History of the Behavioral Sciences* 38 (summer 2002): 259–83; Raymond E. Fancher, "Biographical Origins of Francis Galton's Psychology," *Isis* 74 (June 1983): 228–29; Raymond Fancher, "Francis Galton's African Ethnography and Its Role in the Development of His Psychology," *British Journal for the History of Science* 16 (1983): 67–79; Garland Allen, "Genetics, Eugenics, and Society: Internalists and Externalists in Contemporary History of Science," *Social Studies of Science* 6 (February 1976): 105–22; David N. Livingstone, *Adam's Ancestors: Race, Religion, and the Politics of Human Origins* (Baltimore: Johns Hopkins University Press, 2011); Ivan Hannaford, *Race: The History of an Idea in the West* (Washington, D.C.: Woodrow Wilson Center Press, 1996); Frank Dikötter, "Race Culture: Recent Perspectives on the History of Eugenics," *American Historical Review* 103 (April 1998): 467–78; Diane Paul, *Controlling Human Heredity, 1865 to the Present* (Atlantic Highlands, N.J.: Humanities Press, 1995).

## African Americans, Race, and Eugenics

There are only a few books that have begun to explore the relationship between eugenics, the idea of race, and the lives of African Americans. The historiography has unfortunately ignored the fact that eugenicists devoted considerable resources to the study of black-white differences from the beginning of that movement in the late nineteenth century. Eugenics was not just about preserving whiteness from ethnics, nor was it only social movement; it was also about the construction of scientifically justified color differences. Edward Larson's *Sex, Race, and Science: Eugenics in the Deep South* (Baltimore: Johns Hopkins University Press, 1995) has a short discussion of the effects of eugenic sterilization programs on African Americans. William H. Tucker's *The Science and Politics of Racial Research* (Urbana: University of Illinois Press, 1994) looks at the relationship between eugenics, racial science, and race and the mental health community. As such, the book focuses primarily on the psychometric crusades of the twentieth century and on the influences that psychologists had in the courts in abolishing state-sanctioned segregation. Tucker's latest effort, *The Funding of Scientific Racism: Wickliffe Draper and the Pioneer Fund* (Urbana: University of Illinois Press, 2002), is an important addition to the historiography as it explores the funding mechanisms that helped both eugenic and posteugenic racial science thrive. Gregory Dorr's *Segregation's Science: Eugenics and Society in Virginia* (Charlottesville: University of Virginia Press, 2008) surveys the rise and impact of eugenics on Virginia's racial mores by examining the ways in which eugenics influenced twentieth-century notions of white supremacy in Virginia and how eugenic thinking impacted African American approaches to racial uplift. Dorr's work shows how eugenics quickly became the scientific justification for racial purity in twentieth-century Virginia, and the book is a powerful illustration of how eugenicist ideas about African Americans and about race relations more generally did not simply conform to racial theory but rather shaped it to a very large degree. Lee D. Baker's *From Savage to Negro: Anthropology and the Construction of Race, 1896–1954* (Berkeley: University of California Press, 1998) explores many of the same questions explored in *Race Unmasked* but does so in the context of anthropology. A chapter in Edwin Black's *War Against the Weak: Eugenics and America's Campaign to Create a Master Race* (New York: Four Walls Eight Windows, 2003), 159–82, looks at the career of obstetrician and eugenicist Walter Plecker and his impact on racial and segregation policy in Virginia as the head of the state's Bureau of Vital Statistics. Daylanne K. English's *Unnatural Selections: Eugenics in American Modernism and the Harlem*

*Renaissance* (Chapel Hill: University of North Carolina Press, 2003) offers particularly interesting insight into the relationship between the writings of W. E. B. Du Bois and eugenic ideology. Finally, Mae Ngai's *Impossible Subjects: Illegal Aliens and the Making of Modern America* (Princeton: Princeton University Press, 2004) documents the shift from a scientific and eugenic focus on racial superiority to race difference during the first two decades of the twentieth century.

The Charles Davenport Papers at the APS reveal a striking series of correspondence, lectures, and notes that have generally been ignored by scholars examining the relationship between eugenics and African Americans and the way eugenicists thought about race in a black-white context. In the introduction to *"The Hour of Eugenics": Race, Gender, and Nation in Latin America* (Ithaca: Cornell University Press, 1991) historian Nancy Leys Stepan writes that despite "the historical significance of eugenics . . . it is still surprising how restricted the study of eugenics is, especially when we consider . . . its connections to many of the large themes of modern history" (2). In this context it is therefore not surprising that even given the centrality of racial matters to the eugenics movement the historiography of eugenics and the historiography of racial science have rarely intersected with African American history. Some scholars might argue that this is because race in the context of eugenics was not about the black-white divide in America but about attempts to define whiteness in relation to the immigrants who had been arriving on America's shores since the 1840s. (See, for example, Jacobson, *Whiteness of a Different Color*, 39–90). Susan Reverby's new book on Tuskegee, *Examining Tuskegee: The Infamous Syphilis Study and Its Legacy* (Chapel Hill: University of North Carolina Press, 2009), is an important addition to the historiography of racial medicine that, in part, examines the way scientific thought (including eugenics) enabled the horrors of the Tuskegee Study. See also, for example, James Jones, *Bad Blood: The Tuskegee Syphilis Experiment* (New York: Free Press, 1981); Joseph L. Graves Jr., *The Emperor's New Clothes: Biological Theories of Race at the Millennium* (New Brunswick, N.J.: Rutgers University Press, 2001), 157–72; Gould, *Mismeasure of Man*; N. J. Block and Gerald Dworkin, eds., *The IQ Controversy* (New York: Pantheon, 1976); W. Michael Byrd and Linda A. Clayton, *An American Health Dilemma: A Medical History of African Americans and the Problem of Race*

(New York: Routledge, 2001); Waltraud Ernst and Bernard Harris. *Race, Science and Medicine, 1700–1960*, Studies in the Social History of Medicine (London: Routledge, 1999); George M. Fredrickson, *Racism: A Short History* (Princeton: Princeton University Press, 2002); David Theo Goldberg, *Anatomy of Racism* (Minneapolis: University of Minnesota Press, 1990); Thomas F. Gossett, *Race: The History of an Idea in America* (Dallas: Southern Methodist University Press, 1963); Mark H. Haller, *Eugenics: Hereditarian Attitudes in American Thought* (New Brunswick, N. J.: Rutgers University Press, 1963); Sandra G. Harding, ed., *The "Racial" Economy of Science: Toward a Democratic Future* (Bloomington: Indiana University Press, 1993); Richard Hofstadter, *Social Darwinism in American Thought* (New York: Braziller, 1959); Nancy Ordover, *American Eugenics: Race, Queer Anatomy, and the Science of Nationalism* (Minneapolis: University of Minnesota Press, 2003); Diane B. Paul, *The Politics of Heredity: Essays on Eugenics, Biomedicine, and the Nature-Nurture Debate* (Albany: SUNY Press, 1998); Vanessa N. Gamble, "A Legacy of Distrust: African Americans and Medical Research," *American Journal of Preventive Medicine* 9 (1993): S35–S38; John H Stanfield II, "The Myth of Race and the Human Sciences," *Journal of Negro Education* 64 (1995): 218–31; Lisa Gannett, "Theodosius Dobzhansky and the Genetic Race Concept," *Studies in History and Philosophy of Biological and Biomedical Science* 44 (2013) 250–61; Harriet A. Washington, *Medical Apartheid: The Dark History of Medical Experimentation on Black Americans from Colonial Times to the Present* (New York: Doubleday, 2006).

### The "Rise and Fall" of the Race Concept

Elazar Barkan addresses the historical status of racial science during the twentieth century in *The Retreat of Scientific Racism: Changing Concepts of Race in Britain and the United States Between the World Wars* (New York: Cambridge University Press, 1992). His book argues that through the efforts of progressively minded scientists, scientific racism was by and large pushed out of biological and genetic thought. He contends that scientists in the 1940s took on racial science in true Popperian fashion—falsification. While some mainstream scientists left the idea of race behind, *Race Unmasked* shows that many did not and that racial science remained a force to be reckoned with both because science never divests itself of the biological race concept and because racial science did not need the approval of biologists and geneticists to survive and to thrive. Kenneth Ludmerer also addresses the evolving history of racial science, but only in the context of the evolution of eugenics and its relationship with the emerging field of genetics in a 1969 article titled "American Geneticists and the Eugenics Movement, 1905–1935" (*Journal of the History of Biology* 2 [September 1969]: 337–62). The article later became a chapter in Ludmerer's classic book on the history of genetics, *Genetics and American Society: A Historical Appraisal* (Baltimore: Johns Hopkins Press, 1972). Ludmerer's argument about the relationship between the eugenics movement and the field of genetics, and of the attitude of geneticists toward the race concept is problematic. In his view "new findings of heredity dampened the enthusiasm of many geneticists for the movement; by demonstrating that inheritance is a much more complex process than had previously been thought, these findings indicated to many geneticists that the task of constructing sound and valuable eugenic schemes is not so simple." This awareness was followed by a renunciation of the movement by geneticists "alarmed by the movement's participation in the vitriolic debates over immigration restriction and by its apparent endorsement of the race theories of Nazi Germany" (Ludmerer, "American Geneticists," 338–39). While this was true for many geneticists (including T. H. Morgan and L. C. Dunn) as *Race Unmasked* argues the relationship between eugenics and genetics was never this clear-cut. Specifically, when looking at the relationship between the race concept and the field of genetics, it is clear that the impact of eugenics on genetic thinking outlasted the exodus of geneticists from the eugenics movement. Alexandra Minna Stern's *Eugenic Nation: Faults and Frontiers of Better Breeding in Modern America* (Berkeley: University of California Press, 2005) is the most recent addition to this literature and offers an alternative time line for the decline of eugenics. Stern examines the history of eugenics in California through the 1970s and concludes that it was not until the "protest and liberation movements of the 1960s and 1970s" that the "legacy and longevity of eugenics in the United States" was challenged in a sustained and meaningful way (25).

Published in 2011, Paul Farber's book *Mixing Races: From Scientific Racism to Modern Evolutionary Ideas* (Baltimore: Johns Hopkins University Press, 2011) explores the historical trajectory of the race concept but falls into the same trap as earlier volumes on the subject—it reinforces the "rise and fall" claim that presumes a rejection of the concept by scientists in the post–World War II period. Still, the book does draw attention to Dobzhansky's role in reshaping the race concept in evolutionary biology. Finally, Peggy Pascoe's *What Comes Naturally: Miscegenation Law and the Making of Race in America* (New York: Oxford University Press, 2010) traces the dismantling of miscegenation laws in the United States and how their downfall was shaped by and reshaped notions of race in legal and popular thinking. See also Robin O. Andreasen, "A New Perspective on the Race Debate," *British Journal for the Philosophy of Science* 49 (1998): 199–225; K. Anthony Appiah, "The Uncompleted Argument: Du Bois and the Illusion of Race," *Critical Inquiry* 12 (1985): 21–37; Guido Barbujani, Arianna Magagni, Eric Minch, L. Luca Cavalli-Sforza, "An Apportionment of Human DNA Diversity," *Proceedings of the National Academy of Sciences* 94 (1997): 4516–19; Edward H. Beardsley, "The American Scientist as Social Activist: Franz Boas, Burt G. Wilder, and the Cause of Racial Justice, 1900–1915," *Isis* 64 (1973): 50–66; Juan Comas, " 'Scientific' Racism Again?" *Current Anthropology* 2 (1961): 303–40; W. E. B. Dubois, "Race Traits and Tendencies of the American Negro," *Annals of the American Academy of Political and Social Science* 9 (1897): 127–33; Bentley Glass, "Geneticists Embattled: Their Stand Against Rampant Eugenics and Racism in America During the 1920s and 1930s," *Proceedings of the American Philosophical Society* 30 (1986): 130–54; Julia Liss, "Diasporic Identities: The Science and Politics of Race in the Work of Franz Boas and W. E. B. Du Bois, 1894–1919," *Cultural Anthropology* 13 (1998): 127–66; Lisa Gannett, "The Biological Reification of Race," *British Journal for the Philosophy of Science* 55 (2004): 323–45.

## Anthropology and Race

In *From Savage to Negro: Anthropology and the Construction of Race, 1896–1954* (Berkeley: University of California Press, 1998) anthropologist Lee Baker has covered some of the same ground this book covers from the perspective of anthropology. By examining the anthropological discourse from *Plessy* to *Brown*, Baker argues that anthropology played a primary role in "helping to change the meaning and structure of race for

African Americans" (3). Anthropological discourse is important to the study of the impact of scientific racism, but ultimately it was biologists, and a discussion in the narrower context of genetics, that dictated the ground rules for the geneticization of race and racism and thus the parameters of debates on the nature of race in both scientific and popular discourse during the twentieth century. See also Rachel Caspari, "From Types to Populations: A Century of Race, Physical Anthropology, and the American Anthropological Association," *American Anthropologist* 105 (March 2003): 65–76; Philippa Levine, "Anthropology, Colonialism, and Eugenics," in *The Oxford Handbook of the History of Eugenics*, ed. Alison Bashford and Philippa Levine. 43–61 (New York: Oxford University Press, 2010); Peter Pels, "The Anthropology of Colonialism: Culture, History, and the Emergence of Western Governmentality," *Annual Review of Anthropology* 26 (October 1997): 163–83; Alan H. Goodman, "Why Genes Don't Count (for Racial Differences in Health)," *American Journal of Public Health* 90 (1995): 1699–1702; S. O. Y. Keita, R. A. Kittles, "The Persistence of Racial Thinking and the Myth of Racial Divergence," *American Anthropologist* 99 (1997): 534–44; Jonathan Marks, *Human Biodiversity: Genes, Race, and History* (New York: Aldine, 1995); Audrey Smedley, *Race in North America: Origin and Evolution of a Worldview* (Boulder: Westview Press, 1999); Carol C. Mukhopadhyay and Yolanda T. Moses, "Re-establishing 'Race' in Anthropological Discourse," *American Anthropologist* 99 (September 1997): 517–33.

## Sociobiology

Historians have been surprisingly silent in their exploration of the sociobiology debate. Daniel J. Kevles's *In the Name of Eugenics* and Carl Degler's *In Search of Human Nature* are two major historical works on the topic. Kevles's work includes a brief, dispassionate review of the new synthesis. Degler, however, in the last third of his book, embraces sociobiology, accepting as possible a scientific conception of human nature. Chapter 10 of the present volume was inspired largely by my reaction to Degler's work. His treatment of the emergence of sociobiology fails to interrogate the tenets of the new science, accepting them as the products of a newly advanced objective science. He virtually ignores the many criticisms of sociobiology and sees no connection between sociobiology and its socio-political context. Degler assumes that because "many of the proponents of a recognition of the role of biology in human behavior were and are personally liberal, rather than conservative, in political outlook" and that because "social scientists began to be interested in bringing biology back into the human sciences as early as the 1950s and then through the 1960s, when the political climate can hardly be described as conservative," meant that sociobiological work could not be the product of, or have particular salience in, a specific historical moment (226–27). This assertion is both wrongheaded and troubling. As historian Paula Fass points out in her review of Degler's work in *Reviews in American History*, "In taking sides Degler fails fully to visualize the historicity of the issue, substituting a neo-Hegelian synthesis for the new paradigm that may be required" (240). For overviews of the sociobiology debates from disciplines other than history, see, for example, Ullica Segerstråle, *Defenders of the Truth: The Battle for Science in the Sociobiology Debate and Beyond* (New York: Oxford University Press, 2000). See also Philip Kitcher, *Vaulting Ambition* (Cambridge, Mass.: MIT Press, 1985); Richard M. Lerner, *Final Solutions: Biology, Prejudice, and Genocide* (University Park: Penn State University Press, 1992); S. L. Washburn, "Animal Behavior and Social Anthropology," *Society* (September–October 1977): 35–41; Howard Kaye, *The Social Meaning of Modern Biology* (New Haven: Yale University Press, 1984); Stephen Jay Gould, *Ever Since Darwin: Reflections in Natural History* (New York: Norton, 1977); Arthur L. Caplan, ed., *The Sociobiology Debate: Readings on Ethical and Scientific Issues* (New York: Harper and Row, 1978); Martin Barker, *The New Racism: Conservatives and the Ideology of the Tribe* (Frederick, Md.: Aletheia, 1982).

## Race and Genomics

Troy Duster's *Backdoor to Eugenics* (New York: Routledge, 1990), written at the outset of the Human Genome Project, calls attention to how the technological and philosophical approaches of the then emerging Human Genome Project threatened to reify racial and ethnic constructs in the wake of newly emerging technologies. The book also calls attention to the very fine line that exists between eugenics and genetics, and to how the Human Genome Project could very easily erase any distinction between the two.

Jenny Reardon's *Race to the Finish: Identity and Governance in an Age of Genomics* (Princeton: Princeton University Press, 2005), as mentioned in the introduction to this book, highlights what is a persistent problem in the history of the race concept: the notion of a rise and fall in the race concept in the biological sciences. Reardon believes that this idea is so entrenched that it has become "the canonical narrative of the history of race and science." And I agree with her that as a "dominant narrative," it "truncates history" (22–23). Her book then moves beyond this narrative by showing the persistence of race concepts in the genomic era, with a particular focus on how the Human Genome Diversity Project struggled with concepts of genetic diversity in the 1990s, and the pushback from advocacy groups and research subjects on the group's approach. Reardon's book also helped bring Dobzhansky back into focus as a central character in the history of racial science—an important part of this history that others, including myself, have expanded upon.

Shedding new light on the relationship between race and personalized medicine is the aim of Jonathan Kahn's *Race in a Bottle: The Story of BiDil and Racialized Medicine in a Post-Genomic World* (New York: Columbia University Press, 2013). The book is an interesting exploration of the legal, historical, technological, and market forces that continue to shape a racialized approach to medicine. The book expands our understanding of how the concept of race is utilized in clinical medicine.

Several new volumes have begun to expand the scope and interest in this area, the most interesting of which is Keith Wailoo, Alondra Nelson, and Catherine Lee, eds., *Genetics and the Unsettled Past: The Collision of DNA, Race, and History*(New Brunswick, N.J.: Rutgers University Press, 2012). The collection is packed with essays exploring topics ranging from forensic technology to pharmacogenomics to ancestry testing, and the work in this volume will surely influence future scholars as history continues to turn its attention to this area. Among other new volumes are Catherine Bliss, *Race Decoded: The Genomic Fight for Social Justice* (Palo Alto: Stanford University Press, 2012); Sheldon Krimsky and Kathleen Sloan, eds., *Race and the Genetic Revolution: Science, Myth, and Culture* (New York: Columbia University Press, 2011); Ian Tattersall and Rob DeSalle,

*Race? Debunking a Scientific Myth* (College Station: Texas A&M University Press, 2011); Dorothy Roberts, *Fatal Invention: How Science, Politics, and Big Business Re-create Race in the Twenty-first Century* (New York: New Press, 2012); and Paul Farber and Hamilton Cravens, *Race and Science: Scientific Challenges to Racism in Modern America* (Corvallis: Oregon State University Press, 2009); Evelynn M. Hammonds and Rebecca M. Herzig, eds., *The Nature of Difference: Sciences of Race in the United States from Jefferson to Genomics* (Cambridge, Mass.: MIT Press, 2008); Linda L. McCabe, *DNA: Promise and Peril* (Berkeley: University of California Press, 2008); Barbara A. Koenig, Sandra Soo-Jin Lee, and Sarah S. Richardson, *Revisiting Race in a Genomic Age* (New Brunswick, N.J.: Rutgers University Press, 2008); Lundy Braun, "Race, Ethnicity, and Health: Can Genetics Explain Disparities?" *Perspectives in Biology and Medicine* 45 (2002): 159–74; Morris W. Foster and Richard R. Sharp, "Beyond Race: Toward a Whole Genome Perspective on Human Population and Genetic Variation," *Nature Reviews: Genetics* 5 (2004): 790–96; Lisa Gannett, "Racism and Human Genome Diversity Research: The Ethical Limits of 'Population' Thinking," *Philosophy of Science* 68 (2001): S479–S492; Duana Fullwiley, "The Biologistical Construction of Race: 'Admixture' Technology and the New Genetic Medicine," *Social Studies of Science* (October 2008): 695–735.

# INDEX